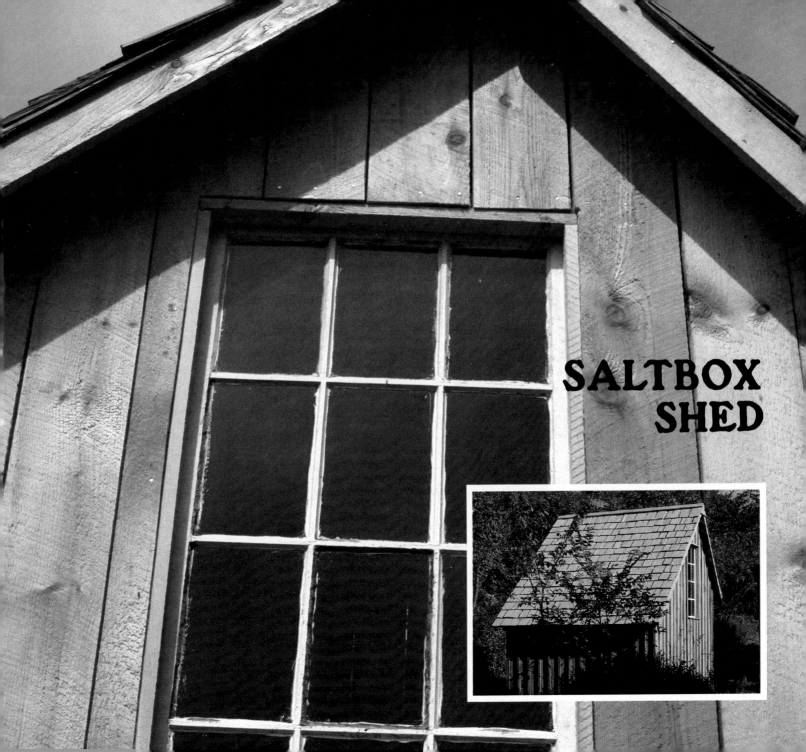

SALTBOX
SHED

THE SMALL
BARN

A POND DOCK

TRUSS BRIDGE

GARDEN
FENCES

OCTAGON
PLANTERS

DROPLEAF HARVEST TABLE

Published first in 1986 by Allen D. Bragdon Publishers Inc.
153 West 82nd Street, New York, New York 10024.

Editor	Allen D. Bragdon
Copy Editor	Karen Ringnalda Altman
Illustration, Photo, Design	by the author

Library of Congress Cataloging-in-Publication Data

Springsteel, Bernard.
   Carpentry & rough wood.

   1. Building. 2. Carpentry. I. Title. II. Title:
Carpentry and rough wood.
TH145.S67 1986   694   86-2646
ISBN 0-916410-33-1
ISBN 0-916410-34-X (pbk.)

Printed and bound in the United States of America

First edition

Carpentry & ROUGH WOOD

How to build seven useful
structures in a country setting

Written & Illustrated by
BERNARD SPRINGSTEEL

Allen D. Bragdon Publishers, Inc.
New York

To Marcia, my wife,
without whom
this book would not
have been possible.

Contents

PREFACE

I love wood. Yet, like a great many other things in our environment, it is threatened. The practice of replanting to keep the ecology and people's needs in balance is a good one, but if you look closely at what's happening, you will see that it is mostly softwoods that are being replenished, because they grow much faster than hardwoods. The softwoods are used mainly for paper products and building purposes. But I worry about the state of hardwoods. They replenish themselves if left on their own—that I can see on my own land. But it takes several hundred years to get a hardwood tree to the dimension needed for cutting. Hardwoods are also being cut down to supply fuel for the increasing number of wood stoves since we have been trying to find an alternative to oil and gas. One wonders how long that can go on. In 1790, far fewer Americans were making wood fires and the number of trees was far greater. I hate to waste a stick of wood when I am building something. Consequently my shop and barn and cellar are full of scraps of wood that I can use at some later date to make smaller items. That makes me feel a little better. The structures I describe in this book are built with softwoods, so I can take some comfort from the fact that the trees I've used replenish themselves at a much quicker rate.

Not only do I like wood, but I also like the way it was used in the early days of our country in both houses and boats. Some of these early designs have held up over the years like "Old Masters," whereas many more recent designs—those 50 or 60 years old—have not held up nearly so well. Wood is something a person can grasp in the hand and shape and feel good about the results. Warming to newer building materials is not one of my fortes, so I want to tell people not to forget about wood. I sometimes pause by a construction site in New York City just to look at the wood used for shoring and for platforms for the cranes, and I think of the different ways that I could use those boards. I often fancy that wood in any form has a life of its own, with feelings and opinions but no choice of how it will be used.

In the barn you see in this book, I am building a boat, and of course I am using a design that was in existence in the 1850s. This "good ship" is being constructed of oak and pine and is much harder to build than a barn. Perhaps my next writing project will be a description of how I built this early fisherman's boat.

A lot of love goes into making something you feel is beautiful. One of the nicest inventions to come along is the camera; it gives one the opportunity to record beauty and pass it on. Wouldn't it be great if someone had had a camera in 500 A.D.?

If I were asked my major reason for writing this book, I would answer that the reason lies in the doing and that special satisfaction one gets from making something by oneself—taking a project all the way from stage A to stage Z. Knowing that one is responsible for the whole picture for better or for worse—in an age of specialization—is a unique experience. So I'm extending an invitation to you to come along and see if what I write and picture will inspire you to begin on an adventure of your own—or at the very least provide you with a good one or two nights' reading in bed.

Ghent, N.Y.
March 1984

King strut truss ready to go up.

Backgrounds

This is not a typical how-to-do-it book; it is more a tale of adventure—an experience that I had and want to share with you. I also hope to impart some of my own woodworking know-how, giving as many specifications as I possibly can, but things change as you go along on any project and ideas come about as you build. You'll inevitably want to make changes to suit your own needs and desires. For example, when I started building this barn, I never intended to put the little greenhouse on it, but my wife, Marcia, kept talking about how she would like to start some seeds in the early spring in a cold flat. So the greenhouse evolved as a large cold frame attached to the barn. This barn combines old and new building techniques and a couple of old architectural styles. To that, add the greenhouse with its modern cantilevered projection.

I wanted to use materials and styles to harmonize with our 1830 Federal home in upper New York state, on 5½ acres of woods and fields and farmland terrain. I had recently discovered logging mills up in that part of the country and was inspired by the full dimensions of the lumber and the price (about 50 to 60 percent cheaper than kiln-dried lumber). Always one for trying to find a way of doing something myself (and economically), I realized that the barn I needed was within fairly easy reach if I built it myself and with rough timber.

Early afternoon shadows over the almost completed barn.

I have always designed things, and in fact it's
what I do for a living, but before I designed the
barn I read a lot of books and looked at many
pictures and paintings of barns. I myself am a
watercolorist and like to poke around barns looking for good subjects to paint. Though books on
how to build things are necessary and useful, ob-
servation is often a better teacher if you have an
eye for details. Classical art students, for example, used to be taught how to paint by copying an
"Old Master." But I find creating something of
one's own that will enhance the environment to
be the most fulfilling aspect of being an artist.

The barn is an artistic and satisfying creation
to me, but it is not technically perfect. It has
curves and slants, and the green logging wood is
beautiful but sometimes has a mind of its own.

I admire the old methods of building because
they were meant for durability and lasting quality.
But because I built this barn pretty much by myself, I did not want to deal with the large timbers
characteristic of the older methods, so I kept the
dimensions within a reasonable and controllable
size. Early structures were built with mortise
and tenon joints. These are very heavy and require a lot of strength and many hands to dupli-
cate. Instead I used galvanized nails and bolts
and oak plugs or pegs for effect (and strengthen-
ing).

TOOLS

Early settlers must have been hardy souls to have done everything with hand tools and to have carved a homeland with these tools. Today we have power tools to help us and save time, but it is great to do some hand toolwork too, just to experience the satisfaction it gives. There is nothing like sliding a plane across a timber to see the shavings fly. In addition to the instant gratification, it gives us an opportunity to reflect on our ancestors and admire their resourcefulness. In this modern age, people are approaching a time when they will again have to be resourceful to overcome new economic, social, and environmental problems.

There are many ways to arrive at the same end and many different tools that will work satisfactorily, but I want to mention how I work and the tools that I favor and why. This barn could be built completely without power tools, but this is not always practical unless you are a real purist—especially when ripping one of those two-by-fours necessary for the corner molding trim.

If you are going to buy tools, try to get industrial quality rather than homeowner quality, lest you end up having to buy tools twice. This is especially important if you plan to use them a great deal, no matter whether carpentry is a vocation or an avocation.

The tool that worked the hardest was the CIRCULAR POWER SAW. Mine was a powerful one, and though it had the smaller 6½″ blade, it could cut through anything up to 2″ thick. Use a good carbide-tipped blade; it will stay sharper than a steel blade. Cutting green timber also tends to burn the tips of steel blades. You may pay twice as much for carbide, but it can be resharpened at least twice professionally, so you end up keeping this blade quite a long time. A combination cross-cut and ripping blade is best. Keep the blade sharp; a dull one will overwork the motor of the saw and shorten its life.

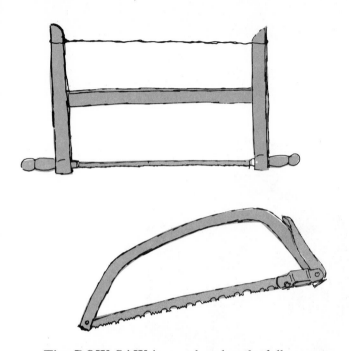

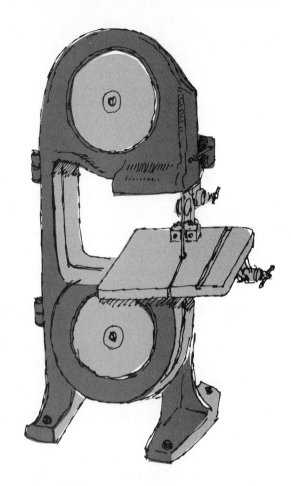

The BOW SAW is another handy fellow, especially for cutting green wood. When you make a mortise cut in the ends of floor beams or in the vertical posts, the circular saw does not go all the way through. The bow saw finishes the job. Designed to cut live trees, the bow saw works very well on green timber. I have pictured here the two types that are available. You can get the newer style in any hardware store, and the older style is available through catalogs from good tool houses.

The BACK SAW usually is a short hand saw that has a thick metal strip along the top of the blade. This keeps the thinner blade from wobbling when you want to make very accurate mortise cuts. BAND SAWS are expensive but great to have. You can make scroll cuts and curved cuts as well as straight or beveled cuts. I have an old used one. For instance, if you are cutting a curved shape in a piece of oak and you have only a hand-power jig saw, you had better have good blades because you will be at the job for some time.

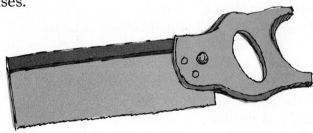

A DADO SAW BLADE can be a combination of two blades joined together or a single blade that adjusts from about ⅛″ to ½″ in width when you want to cut a mortise. It eliminates making two passes with a single blade and chiseling out the waste matter in between.

I used the ROUTER for only one job—making lap joints on the sides of the floor boards and also the barn door slat boards. Again, use only carbide-tipped router blades because the others burn out very quickly. I used a ½″ mortise bit with a ½″ depth to make the lap joints.

The CLAW HAMMER was always with me, hanging from a HAMMER BELT HOLDER, an especially useful article on the roof and anywhere it's hard to stoop down for the hammer every time. Hammers come with two types of claw for pulling nails, one curved and one nearly straight. The curved claw is the more useful on this job because it gives more leverage when pulling nails. And in spite of my fondness for old things, I feel the new steel-shafted hammer is a great improvement over the old wooden-handled one. It has a good feeling of leverage in the hand, and a heavy one gives you maximum driving power for nailing. Hammers come in several weights, but you need a heavy one for those galvanized

25

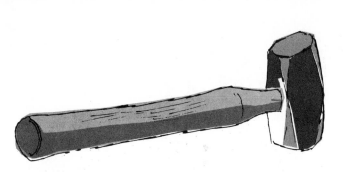

nails. Another handy hammer is the CLUB
HAMMER, used mainly for knocking heavy
boards like roof trusses into place or putting the
wooden pegs in. Its extra weight is needed for
these jobs.

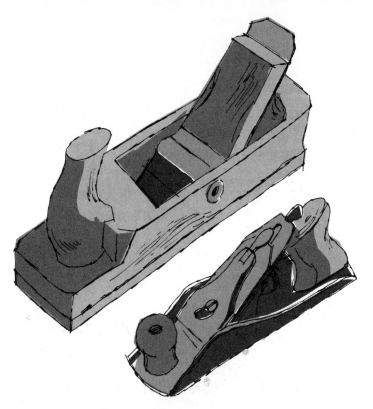

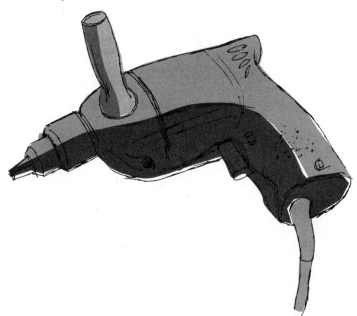

A BENCH PLANE in either metal or wood is
always needed for shaving off that last slice of
wood to get a tight fit.

Of course no job would be complete without a
POWER DRILL. There are a few holes to drill
for the bolts and the wooden pegs.

Large C-CLAMPS and a couple of BAR
CLAMPS (Jorgenson type) stand you in good

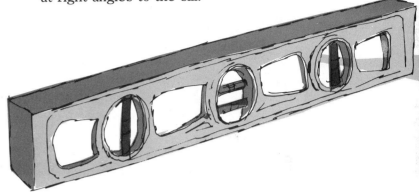

stead when you have only one pair of hands. I used three C-clamps when I was doing the routing for the lap boards. The larger capacity bar clamps come in very handy with the big dimension lumber. You can clamp one end of a board to a post, then get on a ladder with the other end and fasten it.

you a pretty good idea of whether a corner is square or whether you have your vertical studs at right angles to the sill.

A LEVEL that is at least 2' long is an all-important tool from the beginning of the job to the end. It is always important to check to see if your work is straight or vertical.

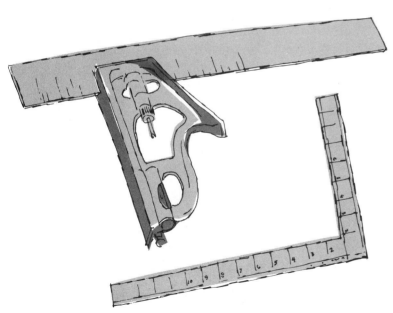

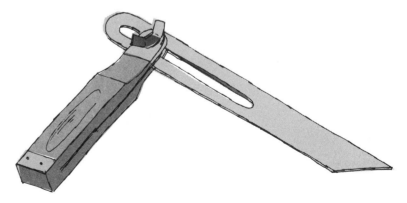

The COMBINATION SQUARE is necessary for straight cuts and checks on whether your cuts come out square. A STEEL SQUARE is a larger flat square with inch markings. It gives

The BEVEL is the only tool I know of to take any angle reading off and transfer it to the piece of wood to be cut.

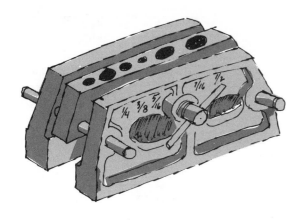

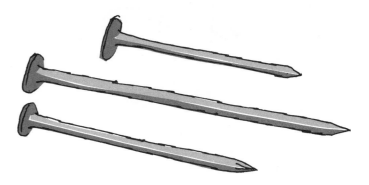

The DRILL BIT GUIDE is a unique tool and can be purchased in several designs. The one pictured here is very simple and automatically clamps the hole to the center of the board. Sizes range from ¼″ to ½″.

All the NAILS for the barn should be galvanized; these last longer against moisture and rusting and are a must for outside work. Also the rough surface of galvanized nails gives them more gripping power. The barn required 3½″ common nails for most of the construction work. For the roof I used a 2″ roofing nail with a bigger head to hold the shakes firmly and a thinner shaft to avoid splitting the shakes. All the bolts and lag screws used were galvanized. Common nails have heads, finishing nails no heads.

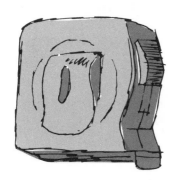

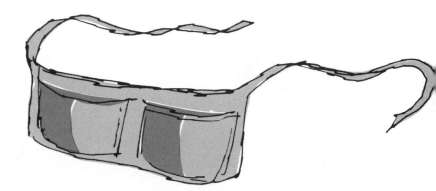

A ROLL-UP STEEL TAPE (10′ to 25′) is the easiest all-around ruler to carry on your person. I find the hardiest type to be one with a brake. You will also need a 50′ roll-up measuring tape for those long measurements involving the foundation and also for comparing your dimensions on the structure from one side to the other.

A CARPENTER'S APRON is invaluable. Usually it has two pockets for two different-size nails, and it's the only way to manage all your paraphernalia when you are hanging off the roof.

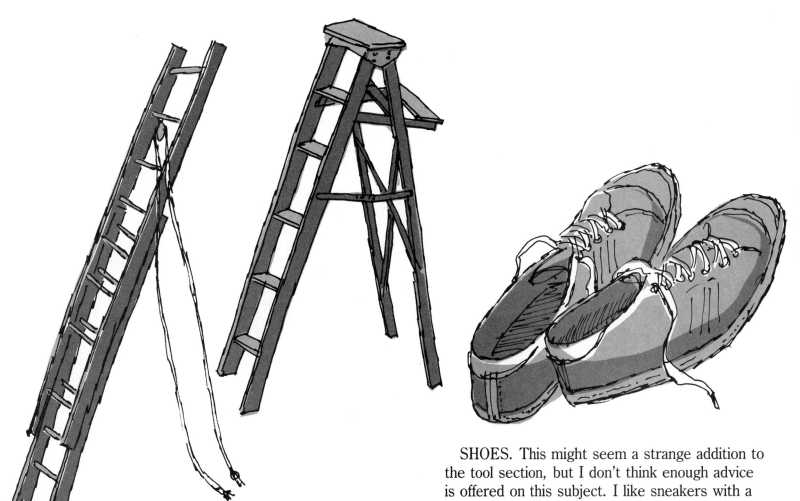

Two different kinds of LADDERS, "movable staircases," figure into this job. An 8' A-LAD-DER was useful for reaching most of the high spots in the barn for up-to-the-top sill work. It can also be used to reach the higher pieces of siding. The EXTENSION LADDER was used nearly exclusively on the roof and on the ends of the building up near the peak. You'll need one that measures about 22' minimum. Aluminum is better than wood unless you feel and look like Charles Atlas.

SHOES. This might seem a strange addition to the tool section, but I don't think enough advice is offered on this subject. I like sneakers with a suction-type base, like boating shoes. When you are crawling around roofs and scaffolding, sneakers give you more of the feel of what you are stepping on than work shoes do. And their lightness and comfort are a blessing. "Work shoes" are for heavy construction and demolition, where you need foot protection.

Terms

I realized in going through my manuscript pages for this book that I was throwing around some terms you may not be familiar with, so here are some definitions just in case.

TOENAIL: putting nails into two members at an angle instead of straight in. It binds the pieces together tighter and forms a strong bond.

SHAKES: wood, as opposed to asbestos or asphalt, shingles. Shakes can be hand split or machine smooth and tapered.

LOGGING TIMBER: green, fresh-cut lumber different from the planed, kiln-dried lumber you buy in a regular lumber yard. It is full-dimension wood that is rough sawn and not smooth. It can be stored on stickers and air-dried. Air-dried lumber is reputed to be more pliable and less brittle than kiln-dried (for this reason it is used to build wooden boats).

PEAK: where the two sides of the roof join at the top.

BEVEL: a degree of angle, which can be transferred from a plan or from a structure with a bevel tool.

RIP: to cut a board through its full length with the grain.

CROSS CUT: the opposite of rip in that you cut across the grain of the board.

GALVANIZED: a process that heavily plates (by dipping in a hot metal bath) steel nails and bolts with a heavy, nonrusting, protective surface of zinc. (Today many stores stock only zinc-plated nails and bolts, and it remains to be seen whether they are as rust resistant as galvanized ones. I'm not convinced that they are.) Galvanized hardware can be purchased at marine supply houses, and at some hardware stores, or directly from manufacturers. A good source is Jamestown Distributors, 28 Narragansett Ave., Jamestown, RI 02835. Send for their free catalog. Very reasonable.

COMMON NAILS have a head on them, whereas FINISHING NAILS have no head or very little. Common nails are for rough construction; finishing nails are for carpentry where you want a minimum amount of nail showing. They can be countersunk and puttied over.

ROOFING NAILS: Nails with a thin shaft but a large flat head that holds the shake down firmly.

SCARFING: a carpentry method that joins two pieces of wood in a strong bond. Page 50 deals with this and illustrates one or two types of scarfs.

STUD: vertical two-by-fours or, in some cases, two-by-sixes, that make up the inner skeleton (walls) of the building. On these rest the top plate and roof.

FLITCH SAWN: a board cut right through from one side of the tree to the other, with the edges left in the shape of the tree. If the tree is large, you can get two boards out of it by splitting the piece in the middle.

BIRD'S-EYE NOTCHES: a 90-degree, two-sided cut in the roof beams. The notch rests on the top sill and braces both the roof beam and the sides of the building.

CUPRINOL: a preservative with a strong chemical ability to prevent wood from rotting. It comes in clear and green.

DADO: an adjustable blade or blades to make a mortise or trench when a board is run through it. You can make a lap joint along a board's edge also.

SILL: the baseboards that rest on the foundation and on which the building rests. They are anchored to the foundation piers or solid cement foundation.

PLATE: the horizontal timbers that rest on the studs and form the resting place for the roof beams. Their edges hold the bird's-eye notches that are cut into the roof beams.

PINTLE: a peg or dowel that sticks up (male end) and accepts the hole (female end); used in this book where the doors are described.

STOP MOLDING: a simple strip molding nailed to the door casing. It stops the door at the proper place, preventing it from swinging through. It also acts as a weather stripping around the door. It can be lined with the soft weather stripping sold in hardware stores.

CASING: the wood nailed to the framing two-by-fours to form a nice frame in which to insert the windows. The facing molding can then be nailed to this.

JOISTS: another name for a roof beam or floor beam.

RAFTERS: also another name for the roof or attic beams.

THE SMALL BARN

The beautiful skeleton—when the structure is the most appealing to me.

Timbers & Foundation

The king strut pegged to the two cross members of a roof truss.

TIMBERS

Logging-timber dimensions are full: 2″ x 4″, 2″ x 6″, 4″ x 4″, 2″ x 8″, 1″ x 4″, and so forth. (If you decide to use kiln-dried lumber with its smaller dimensions, you will need to beef up my specs accordingly. For example, a kiln-dried two-by-four is about 1⅝″ x 3⅝″, and a one-by-six is really ¾″ x 5½″, and so forth.) I also did not have to adhere to the distance between studs that you should have with kiln-dried boards. While I'm on the subject, the floor beams' size and distance apart should reflect what you are going to put on the floor. I would not, for instance, roll a 4000-lb car into this barn without adding half again more floor studs. But I did not plan for it to be a garage.

When purchasing logging wood, it pays to buy a small amount at a time and to PICK YOUR OWN BOARDS! If you simply order a load of lumber, you are likely to get a lot of wood you won't like. I was fortunate in this regard. Where I purchased my wood, they didn't bother me while I picked my boards so long as I left the piles neat. Many lumber yards are chary when it comes to letting you pick the lumber you want.

Timber stacked using "stickers" for ventilation.

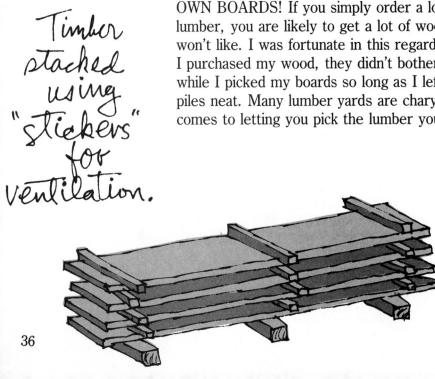

A view of the barn-to-be through the barn-door-to-be. Note window framing and stained shakes drying in the sun.

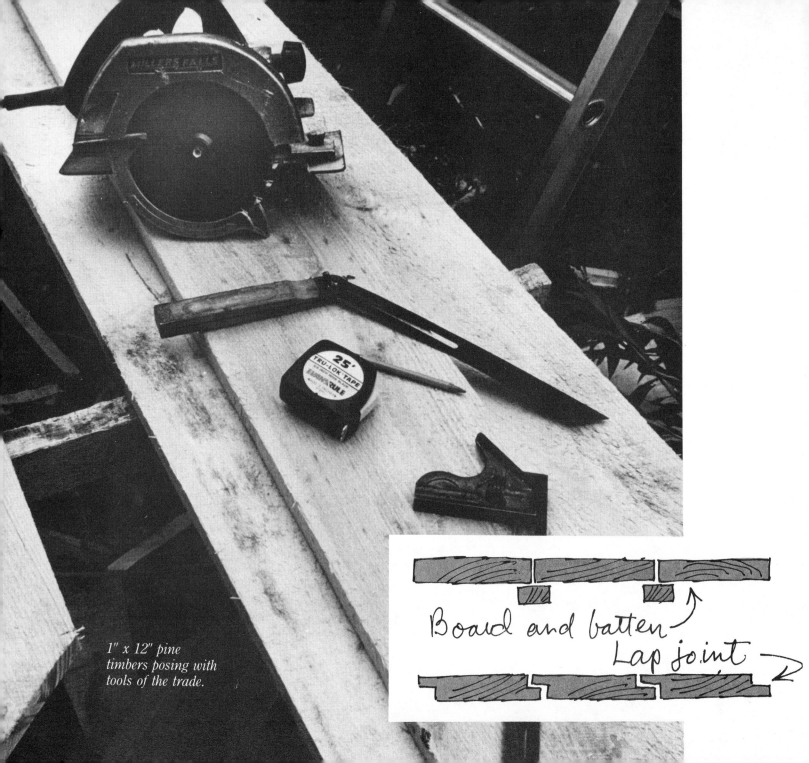

1" x 12" pine timbers posing with tools of the trade.

Board and batten
Lap joint

With the price of lumber today, the least they can do is allow you to select what you want. I took home what the trailer would carry, and I got a good amount of exercise loading and unloading this stuff. You will do well to keep in mind that it is heavier than kiln-dried lumber, so don't overload your vehicle. And a good pair of canvas gloves keeps the splinters in your hands down to a minimum. And put off your lumber shopping if it has just rained, because the boards are about twice as heavy when wet. I was a weekend builder, so I spaced my purchases out. Bring the wood home and lay it on some large timbers supported at at least three points along its length; then put what the trade calls "stickers" between each layer of boards. (Stickers are small pieces of scrap lumber or lath strips.) Whatever you use, try to get them about ¾" to 1" thick and uniform—three to the length of a board. Be sure to pile the different sizes separately so that you can get to the pieces you want. The reason for putting them on the stickers is so that they can dry uniformly. Green wood will warp if the boards are stacked one on top of the other.

The wood also needs air circulation (especially in summer), as unsightly gray fungus will appear when moisture is trapped between the boards. And moisture also encourages rot. (If you're stacking your wood outside, put old boards or a tarp over the top of the pile.) This wood will shrink a bit on its width but none noticeably on its length. If you are planning to butt these boards together, you should know that they will separate somewhat. If you like and want that effect, fine; otherwise you can put narrower batten boards (one-by-twos, approximately) over each seam. This is called the batten board style of siding, which is very well known and very tight. A third alternative is to make a lap joint on each side of the length of board. This will allow a little shrinkage and yet you will still see wood. I used this technique on the floor and doors. I liked the butt look and used it on the triangular wall at each end of the building. Mills cut the following board sizes: 1″ x 4″, 1″ x 6″, 1″ x 8″, 1″ x 10″, 1″ x 12″, 2″ x 4″, 2″ x 6″, 2″ x 8″, 2″ x 10″, 4″ x 4″, and 6″ x 6″. They will cut special sizes to order.

Little attention is paid to grades of lumber nowadays, except to distinguish clear from common; perhaps there should be a third grade for rag-tag pieces, priced accordingly. It's great fun to go to the logging yard, as you get to feel the heft of the boards and you learn about the different quality of the pieces. Making friends with the proprietor is a good idea—it leads to chatting about wood and trading knowledge. This kind of encounter can make you feel better about the world.

FOUNDATION

The first step in building is to decide on the type of foundation. I elected to put the barn on six piers. This meant no span of wood in any direction was greater than about 11½', since the barn is roughly 12' x 24' finished. Piers, compared with a full foundation, have a rustic appearance. Stones that are flat in shape make a beautiful foundation pier when cemented together. Some barns supported with flat stones laid one on top of the other without cement have withstood many winters. Piers also allow air to circulate under the building, which keeps everything dry. Wood's worst enemy is trapped moisture, which in turn encourages termites that love the darkness and dampness. And eventually wood rots if it never really dries out.

It is normal to put the pier or footing down below the frost line, usually anywhere from 3' x 5' (except in the South). I hit shale rock, which is very common in my area, at this point in my diggings. I built my piers right on the shale when I was sure I had hit a concentrated mass of it. The piers can be made by building forms which you then fill with concrete or by stacking 8" x 8" x 16" cement blocks, which you may also fill with concrete. It is wise to put sheet metal on top of the piers before resting the wood on them. The metal is helpful in discouraging termites. (It is more important, though, to check the piers now and again to make sure the little bruisers haven't made one of their mud tunnels up to the wood.) Anchor the barn to the piers with anchor bolts, nuts, and washers that can be purchased at a regular lumber yard. Simply sink the right-angle end into the wet cement, leaving enough bolt showing to allow for the thickness of the angle iron, washer, and nut. Use heavy angle irons with holes for the lag screws and anchor bolt. Two lag screws fasten the iron to the 4" x 4" post.

Before you can place the piers, you must lay out the dimensions of the barn and locate the piers' positions. The illustration shows how you can create the rectangular shape of the structure with stakes and line. It's best to put the stakes beyond the hole digging area. Then construct a large wooden triangle (6' x 8' x 10'). This will give you a true right-angle corner. A carpenter's metal angle square is not big enough, so you can

Framing out the bones of the attached greenhouse.

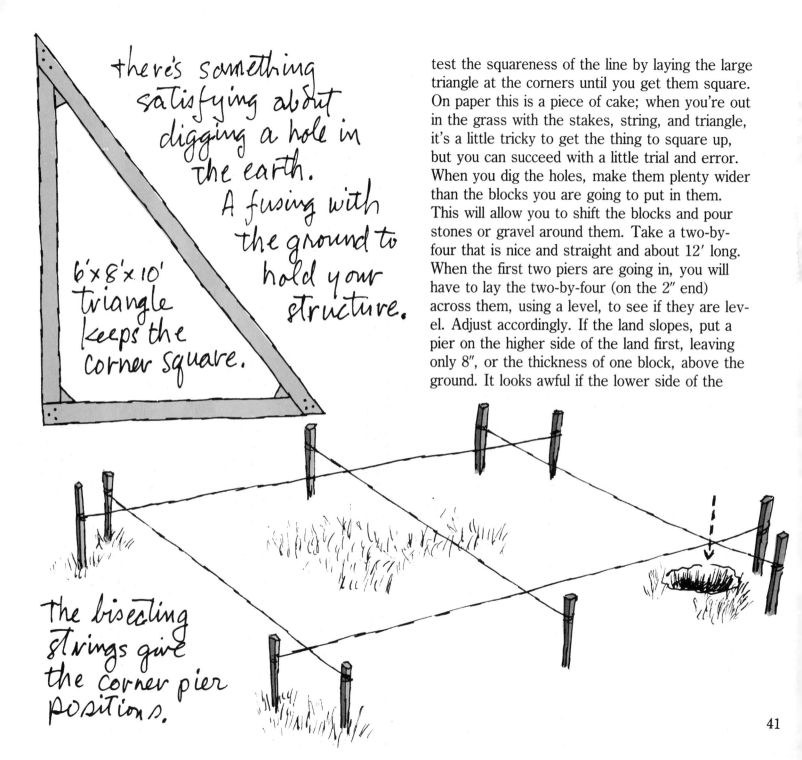

there's something satisfying about digging a hole in the earth.
A fusing with the ground to hold your structure.

6' x 8' x 10' triangle keeps the corner square.

The bisecting strings give the corner pier positions.

test the squareness of the line by laying the large triangle at the corners until you get them square. On paper this is a piece of cake; when you're out in the grass with the stakes, string, and triangle, it's a little tricky to get the thing to square up, but you can succeed with a little trial and error. When you dig the holes, make them plenty wider than the blocks you are going to put in them. This will allow you to shift the blocks and pour stones or gravel around them. Take a two-by-four that is nice and straight and about 12' long. When the first two piers are going in, you will have to lay the two-by-four (on the 2" end) across them, using a level, to see if they are level. Adjust accordingly. If the land slopes, put a pier on the higher side of the land first, leaving only 8", or the thickness of one block, above the ground. It looks awful if the lower side of the building is too steep (more than 3'), unless you are making a full stone foundation.

You may find that one pier is slightly shorter than it should be compared to its opposite pier. You can build up the difference with bricks and cement or a 4" x 8" x 16" block. The advantage to using wooden forms for cement piers is that you can build all the forms and set them in the ground and check them for levelness prior to pouring the cement. A very popular form in current use is a round cardboard tube which makes a neat cylindrical pier.

I thought this part of the building program would be a snap, but I was wrong. Go carefully and make sure that the piers come out level; otherwise you will have a crooked barn. The land sloped a bit on my project, so naturally I had to figure the piers on the upside to be shorter than those on the downside. If you need to make a similar adjustment, this is an important precalculation. Set up some cement blocks near the hole and run the two-by-four and level on them to see how high the blocks will stick up on the downside. Mine ended up being about 2½ blocks high, or about 20" compared to the 8" (one block) on the upside.

Now you may have noticed some barns that seem very close to the ground. This condition makes me very nervous, even though I know barns can take a lot of punishment and still remain standing for years. But being close to the ground means that dirt can build up around the wooden sill, and rot is almost bound to set in, even with preservatives in the wood—not to mention the probability of visits from our old friend the termite. So get it up high enough to

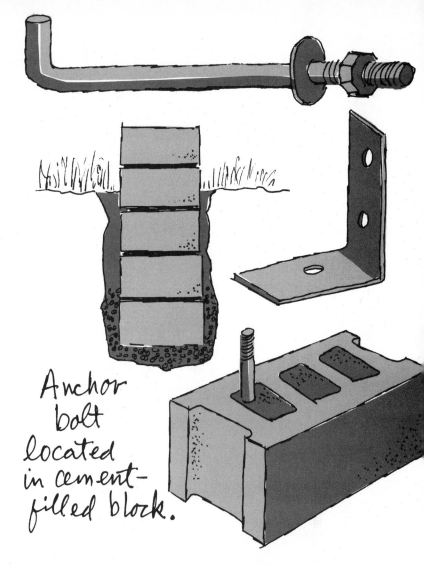

Anchor bolt located in cement-filled block.

see underneath and across to the other side. You can also inspect your barn by crawling under there once in a while to see how the beams are holding up. The wood sill (wood that rests on the piers and supports the bulk of the building) should be treated to a bath (try twice) of that good green Cuprinol. It looks awful but lightens in time and can be stained or painted over.

The angled staircase gave the back of the barn an interesting touch.

Building the Bones

THE SILL

I think pictures tell a lot more about structures than words do, so I'm going to try to give you precise visuals and label the wood sizes. My sill is built up of three pieces of wood: a two-by-six on its 2″ side all around and 2 two-by-fours stacked inside in an "L" shape. This design provides support strength and gives the floor beam two-by-eights a notched position to sit on. The sill should be put together first, then tested using the straight two-by-four across with the level. Do not be surprised to find you're not level. The planets aren't perfect and neither is timber lumber. In the places where you want to add some height, use some pine scrap or red cedar shakes soaked in Cuprinol to shim under the two-by-sixes where they rest on the piers.

Note on the drawing how I connected the 2 two-by-sixes where they meet over the middle pier. Since it is just about impossible to get timbers 24′ long anymore unless you cut your own, it is more practical to take 2 two-by-six-by-twelves and make a joint called a scarf. This method is much better than butting the two pieces, as there is no real strength in butting.

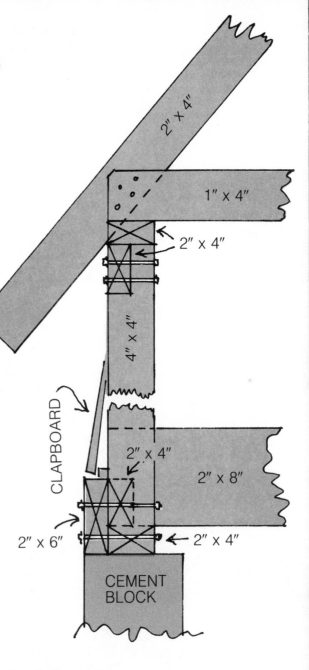

2″ × 4″

1″ × 4″

2″ × 4″

4″ × 4″

CLAPBOARD

2″ × 4″

2″ × 8″

2″ × 6″

2″ × 4″

CEMENT BLOCK

The small barn in its very early stages. The next steps are the studs and framing out of the doors and windows.

Saturate the joints with Cuprinol before and after joining. The two pieces of wood are held with lag screws (predrilled) and washers—one from above and one from below.

CORNER POSTS AND JOISTS

The four corner posts and two center posts (four-by-fours) on the long sides rest on the six piers. These each have two stove (or machine) bolts, washers, and nuts let into the 2″ x 6″ sill boards. Besides the bolts, I also toenailed several 3½″ galvanized common nails from the two-by-sixes into the four-by-fours. Toenailing is nailing at an angle rather than perpendicular. This is a stronger nailing technique and forms a locking pattern. These four-by-fours are about 10′ long, and I notched (2″ x 4″) the tops before erecting them. Hold the four-by-fours against the 2″ x 6″ sill with a big C-clamp or Jorgenson-type clamp; then use the level to get them relatively straight. Drill the two holes and the four-by-fours together using ⅜″ or ½″ gauge bolts. Do not drive any nails until after you insert the bolts because your drill bit might hit them. When the 6 four-by-fours are in place, they must then be joined to each other at the top plate. This is done with two-by-

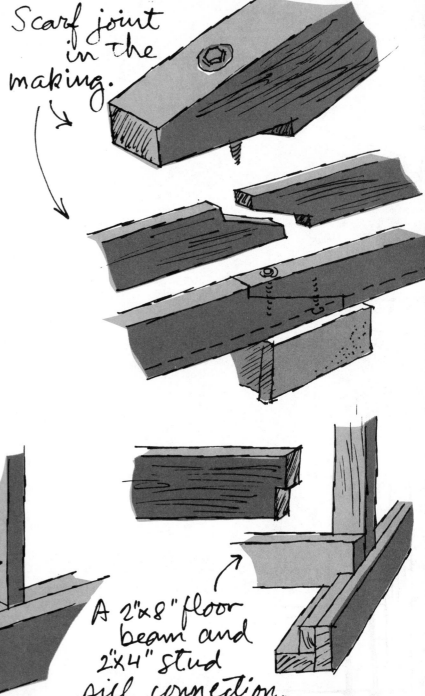

Scarf joint in the making.

Toenail

A 2″x8″ floor beam and 2″x4″ stud sill connection.

46

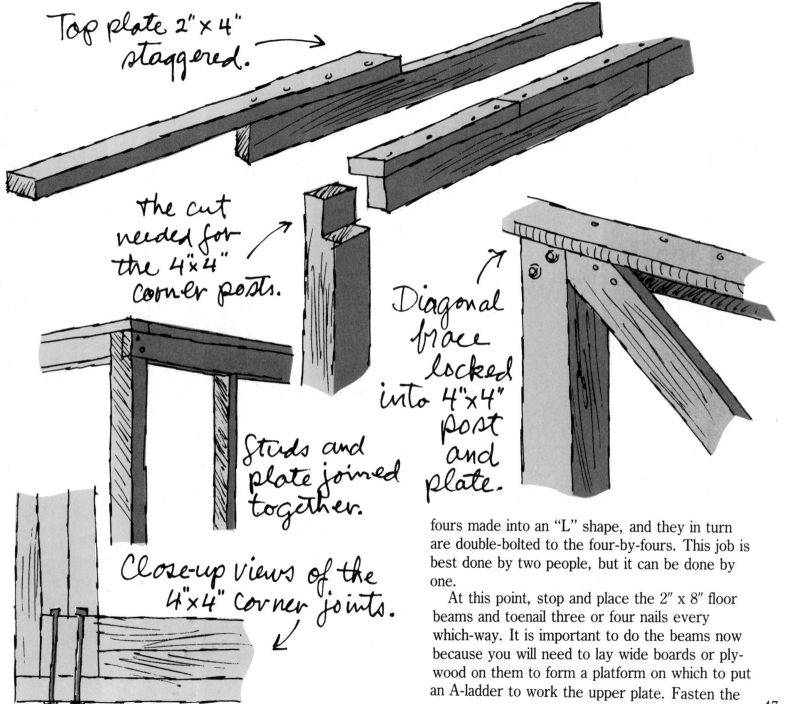

Top plate 2" x 4" staggered.

the cut needed for the 4"x 4" corner posts.

Diagonal brace locked into 4"x4" post and plate.

Studs and plate joined together.

Close-up views of the 4"x4" corner joints.

fours made into an "L" shape, and they in turn are double-bolted to the four-by-fours. This job is best done by two people, but it can be done by one.

At this point, stop and place the 2" x 8" floor beams and toenail three or four nails every which-way. It is important to do the beams now because you will need to lay wide boards or plywood on them to form a platform on which to put an A-ladder to work the upper plate. Fasten the

47

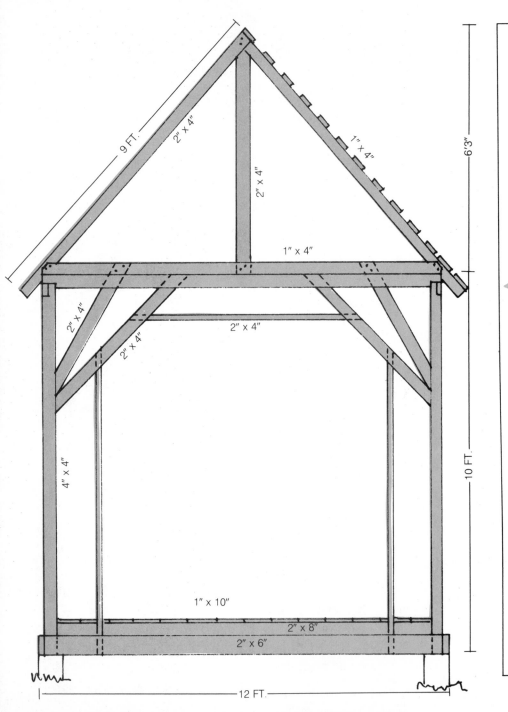

9 FT.

2" x 4"

1" x 4"

2" x 4"

1" x 4"

6'3"

2" x 4"

2" x 4"

2" x 4"

4" x 4"

10 FT.

1" x 10"

2" x 8"

2" x 6"

12 FT.

In early times mortise and tenon joints kept the heavy timbers together.

A dowel was used to lock a tenon.

boards or plywood lightly before you use this platform, though; otherwise you may make the dangerous mistake of stepping on the wrong end, flipping the board, the ladder, and you.

Back to the top plate. In the illustration you can see how the two-by-fours are staggered for more strength. To do this job alone, first fasten 2 two-by-fours. Then climb the ladder and lay this unit on the middle post first. Take your hammer and a 3½″ nail with you to fasten it temporarily. Do not strike the nail all the way home as you will probably want to remove it for adjustment. With the plate thus fixed, you can then scurry down the ladder and move to the end post where the other end of the "L" is waiting for you. This can be fastened accurately because it is the right length to fit the end corner. Test the vertical accuracy of the two end posts to see if the middle one needs adjusting. (If you think you can nail this whole piece up in one shot—don't. Even if you have Hercules at the other end helping you, it will probably sag in the middle during the lift and get out of whack.) It is then a very simple matter to connect the other missing two-by-fours one at a time. (TRICK: Presaw all the pieces on the ground; then you can fit them with the knowledge that they will be reasonably accurate in length.) The two end precut crosspieces should then be connected. I found the best way to do this is to put the ladder in the middle, then carry the two-by-four up and plop it on each end post. Keep a wary eye on this member as you run to nail a corner in case it falls unluckily; nail one end, then the other. Put a Jorgenson clamp

2″ x 8″ joists are in. Note the 1″ x 10″ floor boards for standing the A-ladder on.

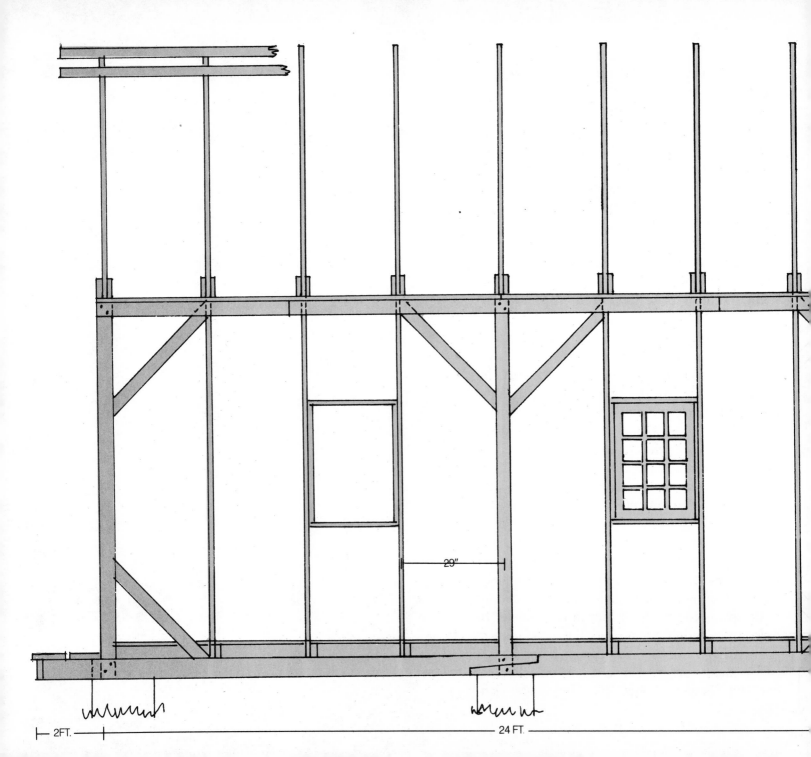

29"

2 FT.

24 FT.

the diagonal braces that stiffen the structure also became an architectural detail that set an early American style.

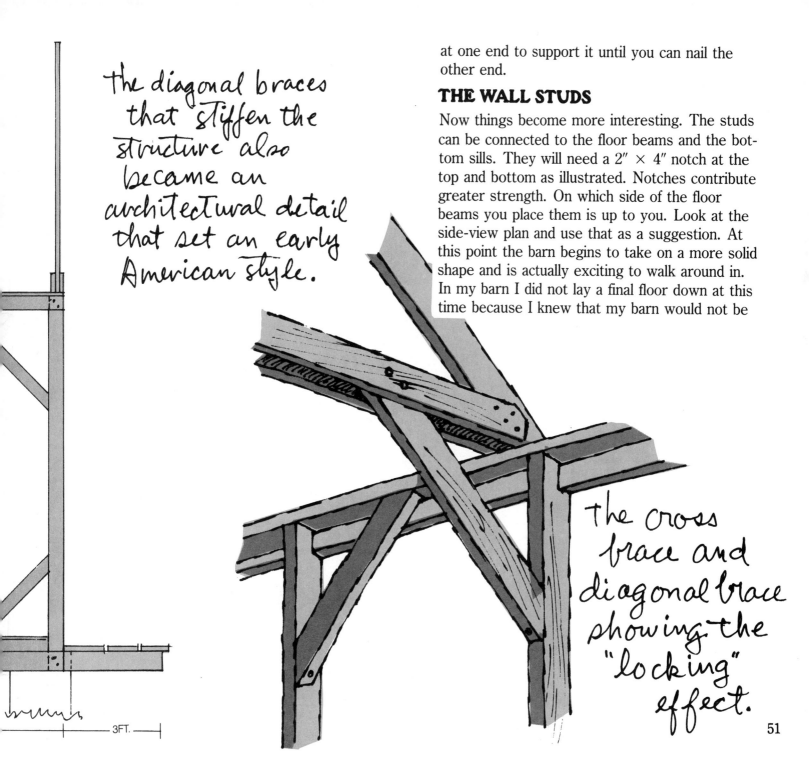

3FT.

THE WALL STUDS

at one end to support it until you can nail the other end.

Now things become more interesting. The studs can be connected to the floor beams and the bottom sills. They will need a 2″ × 4″ notch at the top and bottom as illustrated. Notches contribute greater strength. On which side of the floor beams you place them is up to you. Look at the side-view plan and use that as a suggestion. At this point the barn begins to take on a more solid shape and is actually exciting to walk around in. In my barn I did not lay a final floor down at this time because I knew that my barn would not be

the cross brace and diagonal brace showing the "locking" effect.

51

finished for a while, and I didn't want to expose the floor to the weather.

DIAGONAL BRACES

These delicious pieces of wood cut at 45-degree angles act as stiffening agents and really brace the whole box together. First I put them on the top, but then I realized the bottom could use them too. They look terrific and can be pegged as well as nailed. In the old days, the braces were tenoned and mortised into the plate and vertical post and formed a solid bracing network for these structures. Pegged and nailed, they do a similar job. Thus the braces have a practical as well as an aesthetic reason for being.

ROOF TRUSSES

Now the going gets tough. If you have neither a willing friend nor a strapping (and also willing) 16-year-old son, you should hire a part-time helper.

PEGS

This is a good time to talk about pegs. I took scrap oak ¾" thick and cut pieces ¾" x ¾". I put the pieces in a vise and with a small plane (or spokeshave) made the pegs slightly rounded but not perfectly so. This gives them more holding power when you drive them into a round hole. (Drill ¾" holes.) The pegs look better protruding out on each side about ¼" and also hold better. You can soak them in water if you like so they will swell, but I did not find this necessary.

I put up the first roof truss myself, and I want you to know that it was a hair-raising experience not to be repeated. At worst I could have let it topple to the ground, but that would have hurt my pride, not to mention the truss. Fortunately, my new-found friend and weekend guest Spiro appeared at this time and offered to help with the roof raising.

I built the nine trusses on the ground first. Four of them had king struts and five did not. I felt four were enough, and I liked the look of the alternating pattern. You might consider doing it this way especially if you want a platform upstairs in the barn, because a king strut for every truss will get in the way. These struts were nailed and pegged. I had to use 5" nails (one on each side) to fasten the king struts to the two-by-fours. Predrill holes to avoid having the wood split. There are many ways to do these trusses. The idea of the 2 two-by-fours on either side of the roof beam appealed to me because it allows a 2" diagonal brace into the space provided and thereby attaches the roof trusses to the lower frame of the barn—locking on the roof, as it were. Although I felt a lot safer after those braces were in place, it wasn't until the roof was finished that I felt genuinely proud.

Once the scaffolding is up and the nine trusses finished, you are ready for a truly exciting moment —the roof raising! Spiro and I would pick up a truss and hang it upside down between the upper sills, as pictured. Then—one of us on a ladder, the other on the scaffold—we would get a firm grip on the roof beams and swing the truss into an upright position. (With two people, the weight of the truss is controllable.) Then we toe-

Four of the roof trusses ready to be put up.

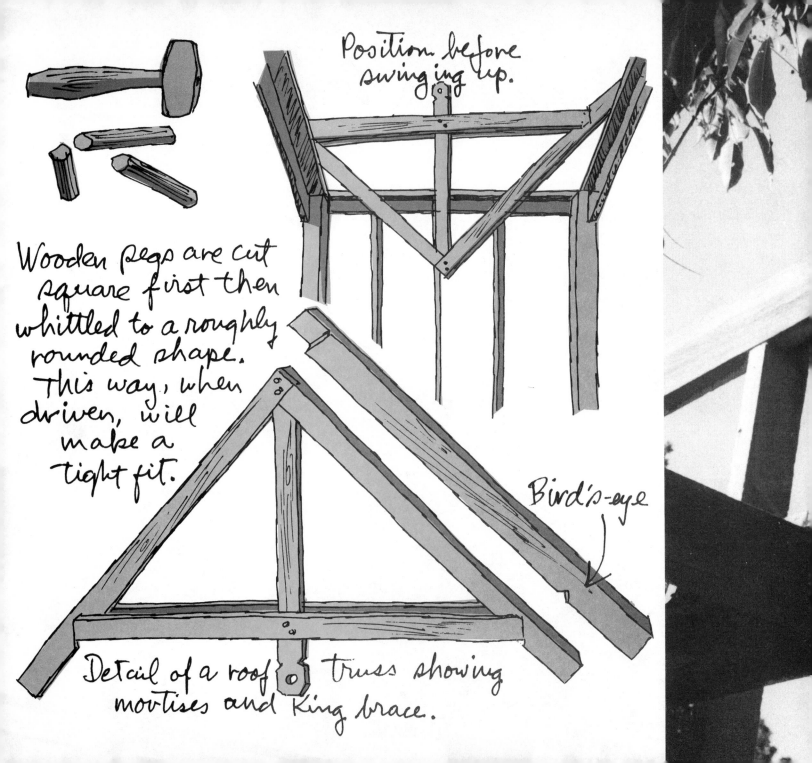

Position before swinging up.

Wooden pegs are cut square first then whittled to a roughly rounded shape. This way, when driven, will make a tight fit.

Bird's-eye

Detail of a roof truss showing mortises and King brace.

The first roof truss in place— the one I put up myself that almost went sailing.

nailed it to the sill on each side. (NOTE: The bird's-eye notches should be the same distance apart as the width of the building. They are a little wobbly at this point but nothing to worry about unless a hurricane is brewing.) The nine trusses go up pretty easily in an afternoon. Before you go to dinner, put two rows of one-by-fours across the beams that are the roof slats. This will hold all the beams together tightly until you are ready to put more roof slats on. When I

A WORD ABOUT SCAFFOLDING

You can buy scaffolding, but it will be cheaper to make your own, and you can use the wood later on in different parts of the building. The illustration shows the way I did it and it worked fine. (When you build something yourself and know you're going to have to depend on its strength, you really have second thoughts. One of mine was to realize how many of the things we use in everyday life are manufactured for us and we assume that they are safe merely because someone else made them.) I liked this new experience of designing and building something I had to entrust my safety to, though. It certainly gave me an appreciation of the faith people over the ages have had in themselves and their ideas. Anyhow, you can build this scaffold one side at a time. Use lots of nails. Toenail at least four nails at joints and from both sides of the timbers. The floor of the scaffolding should put your hips at a level with the roof edge. To get the roof trusses up, one person can stand on the scaffold while the other stands on an extension ladder against the other side of the structure.

the scaffolding.

reached this point, I turned to the job of putting up those 2″ x 4″ diagonal braces that tie the roof trusses onto the frame of the barn. As you can see in the photo, the roof trusses have bird's-eye notches that clamp over the upper plate, holding the building together. The diagonal braces are nailed to hold them in position and then pegged.

The 1″ x 4″ roof slats can be nailed all the way to the top on one side. They serve as a ladder as you climb toward the top. This is a very heady moment in the building of your barn, because you

"Raise high the roof beams" is very applicable here. The roof trusses are up and the 1" x 4" slats completed on one side.

TIMBER SHOPPING LIST FOR BARN

This is an approximate timber list but you will need more 2″ × 4″ and 1″ × 4″ pieces that crop up all over. There is no listing for timber for the Scaffolding but I used a lot of 2″ × 4″ lengths that later ended up in the barn. Greenhouse list on page 80. All wood is pine except where noted.

SILL AND FLOOR BEAMS
Three 2″ × 6″ × 16′
Three 2″ × 6″ × 12′
Ten 2″ × 4″ × 12′
Nine 2″ × 8″ × 12′

FLOORS
Twenty-eight 1″ × 10″ × 12′

STUDS
Six 4″ × 4″ × 10′
Twenty-one 2″ × 4″ × 10′

PLATE
Twelve 2″ × 4″ × 12′

DIAGONALS
Six 2″ × 4″ × 10′ (side braces).
Three 2″ × 4″ × 12′ (cross braces).

DOOR FRAMING
Four 2″ × 4″ × 14′

WINDOWS
Three 2″ × 4″ × 12′
Three 1″ × 4″ × 12′

FRONT DOORS
Four 1″ × 4″ × 10′ (trim strapping)
Eight 1″ × 4″ × 8′ (trim strapping)
Two 1″ × 8″ × 14′ (facing)
Six 1″ × 8″ × 12′ (facing)

BACK DOOR
Three 1″ × 8″ × 8′ (facing)
One 1″ × 6″ × 14′ ("Z" brace)

FRONT PORCH
One 2″ × 6″ × 12′
Two 1″ × 12″ × 12′

BACK PORCH AND STEPS
One 1″ × 12″ × 10′
One 2″ × 8″ × 12′
One 2″ × 6″ × 12′ (sill piece and beams)

ROOF
Eighteen 1″ × 4″ × 12′ (horizontal straps)
Eighteen 2″ × 4″ × 10′ (rafters)
Two 2″ × 4″ × 14′ (king struts)
Forty-four 1″ × 4″ × 12′ (slats)
Eight 1″ × 6″ × 12′ (slats)

DOOR HINGES
One 1¼″ × 5″ × 10′ OAK

CORNER MOLDINGS
Eight 2″ × 4″ × 10′

CAP
Four 1″ × 6″ × 12′

ROOF-END FACING
Four 1″ × 4″ × 10′

ROOF SHAKES
Five SQUARES (5 bundles each) (one square covers approximately 100 sq. ft.)
Three BUNDLES of smooth shakes

SIDING CLAPBOARDS
Sixty-four ¾″ × 8″ × 12′
Sixteen rows ("flitch" sawn) approx. 8″ to 9″ wide.

VERTICAL SIDING
Four 1″ × 8″ × 10′
Four 1″ × 8″ × 14′

FOUNDATION
Thirty 8″ × 8″ × 16″ cement blocks
Five 80 lb. bags concrete mix
Six 12″ anchor bolts, washers and nuts.

1"x4" slats are every 9 ft. on center.

First course of shakes is laid double to seal openings.

Detail of roof beams, slats and corner 4"x4" studs.

can look out upon the wonders of nature as well as appreciate your own ability to get a rooftop about 17' from the ground. Be sure to stagger the lengths of the 1" x 4" butts. This gives greater strength to the structure. And be sure to get about 8" to 10" of overhang with those slats at either end of the roof. (Some buildings are made without any overhang, like a Cape Cod saltbox.) The reason I chose to finish one side of the roof at a time was to avoid moving the scaffold. You need the scaffold mainly as a place to put the split shakes on and to stand on to work the lower shake courses (rows). Note the two bottom rows of slats are one-by-sixes instead of one-by-fours because they overhang and I wanted more firmness there.

*Best view of the scaffolding in use and
the ladder at angle for
roofer to stand on.*

Detail of shakes, charmingly uneven yet straight.

SHAKES

Two basic types of wood shakes are carried by lumber yards: the tapered sawn type, which are relatively smooth, and the type I used, which are called hand splits. These are about 24″ long, whereas the smooth ones are about 16″ long. The roof of this barn is about 480 sq. ft., so I used five squares of shakes. A square is comprised of five bundles, and a bundle covers about 20 sq. ft. if you leave about 9″ of shake showing to the sun. Because split shakes are bulky and thus do not lay well, any exposure less than 9″ is not very practical. If you prefer a tighter look with less shake showing, use the smooth ones. In fact, 9″ of exposed surface would be too much for the smooth shakes; 5″ or 6″ would be more

appropriate.

Start at the bottom edge; the first row must be doubled in order to prevent gaps. I used one bundle of smooth shakes for the first course to provide a flat base for the splits. The shakes are made from red cedar and are very rot resistant. When applying the shakes, you have to overlap the joints so that the rain will not be able to penetrate. Use 2″ galvanized roofing nails for the splits because they are very irregular in thickness. Put only two nails per shake about 1″ in from either side just above the line for the next row so that they will be covered by that row.

You'll love this! I dipped each shake in a 3-gallon pail of stain; then with a brush I sloshed the stain two-thirds of the way up the outside of each shake and left them all to dry before using them. This is a nice way to preserve the shakes and get stain on the hidden surface, but it can be tiresome. You may prefer to leave them natural, which is fine because they are tough and durable, or you can stain them on the roof. Don't try a roller on a pole—it is not an effective method for covering shakes because their surface areas are too irregular. The sloshing-in-a-bucket method is really the most effective and easiest way in the long run.

If you can, shop around for these shakes as they can vary in price. As you can see from the illustration, the 1″ x 4″ slats are spaced so that the rows will have something to nail into at the proper height. It works like this: if you have 9″ of shake showing, then the one-by-fours should be spaced 9″ apart on center. Easy.

In shingling I was able to nail the first five or

Shake staining.

six rows while standing on the scaffold (it depends how tall you are). At some point you will not be able to reach high enough, so you must get up on the roof. At first I purchased a pair of roofers' angle braces, which—in theory—you nail to the roof you have already completed and lay a 2′ x 10′ board across them so you can kneel on this platform and work. This proved to be an unacceptable type of aid to use with these shingles, which are very thick and have an irregular surface. (It works only on asphalt shingle roofs.) roof slopes at a little less than a 45-degree angle, which makes it pretty steep. So my neighbor Vern and I (it's best to have a partner at this point) realized the best way to solve our problem was to extend our 24′ extension ladder as far as it would go. We then laid it against the roof along the pitch. At first this seemed precarious. The angle of the ladder will make it extend quite a way out from the side of the house. But have faith—these ladders can take it. Now one of us could walk up the ladder from the ground all the way up the roof to do the nailing. The other could get off on the scaffold, pick up an armful of shakes of different widths, then follow the nailer up the roof where he could hand him the shakes as he needed them. If you had to do this alone, you would be going up and down a lot, which increases both fatigue and risk. NOTE: Since the shakes come in a random variety of widths, when

laid randomly the effect is more charming than if their widths were uniform. The gaps between the shakes in each course must be covered by a shake in the course above it, naturally, so you will be in for a lot of eyeball measuring to select a wide enough shake to cover the next gap, or a narrow enough one to fall at least 1″ short of it so the next one will cover. Otherwise you will have to split some shakes with a hatchet.

You can store some of the shakes between the 1″ x 4″ roof slats and use them as you need them. The work goes faster if the lower fellow looks for the right widths and splits some, if necessary, on the scaffold. Start with the ladder near the left-hand end of the roof. Nail all the rows partway along as far as you can reach without moving the ladder. Start at the bottom and go all the way up to the ridge line, then move the ladder along to the right. There is no need to finish a course as you did on the scaffold where you could walk the length of the roof. Some carpenters might say you have to mark each course with a chalk line, but with hand split shakes all you need is this little device that fits in your back pocket or apron: cut little straight-edge sticks out of wood about 9″ long. Every two or three shakes lay the stick along the edge of the lower

the staggered cap boards for the roof peak.

row to see if you are going along straight. Most people *want* a little up and downness with these split shakes to achieve a rustic look, but with this method the row itself will never meander or look crooked. We were able to do as many rows in a section as we wanted with this little guide. Of course, if you were to use all smooth wood shakes, which are neater, it might be advisable to lay a string from end to end or take a 1" x 2" batten and nail it 6" above the course below. You move it up 6" for each course and it gives you something to lay the edge of the shake against. Put little protruding nails in the batten stick so you can pull it out and move it. The batten should be straight, with no big knots, and about 8' to 10' long. Another common method is to use string stretched the length of the roof and fastened at either end with a nail, but I find the string too flexible and to me it's a real pain to travel the length of the roof each row to move the holding nail up to the next row.

When at last you reach the top, you can take a breath and look at the view for the second time and marvel at how much easier it is for nature to produce things than for people to do so. The last two rows will need cutting on the length. Measure how much and have your partner cut them on the ground and bring them up. On the last row, do not be tempted—as I was—to leave it a little longer than 9" because of the cap. It gives the effect that something is missing. It's better to have a stubby little row with a cap, because it will look more uniform from the ground.

THE CAP

After both sides of the roof were shingled, I made the cap from one-by-sixes. I cut one to 5" in order to allow for the 1" overlap at the top, so I ended up with an equal 6" overlap on each side. I was going to do this piecemeal, but I sat down on a log and thought about it a lot. My son Peter, who likes to build, was with me that weekend and I knew he liked to straddle rooftops, so I decided to build the cap on the ground, then drag it up there and plop it on. One advantage of this method is that it is in one piece when you finish building it, so you can lay it on the peak despite the ups and downs of the irregular shakes. On the ground you are also able to stain it topside and underneath for better lasting quality.

I used the same method to make the cap that I used to make the upper sill. If you remember, the pieces were staggered for strength and the butts were separated. Carefully measure the length of the roof's ridge to make the cap the right length. My son scurried along the ridge very nicely to do this. The measurements on the plans do not count anymore, because the real thing changes size here and there as you build. With a stout heart, you cut it to the right length on the ground and stain the ends as well. You are now ready to go airborne, which is much easier than you might think. Together we lifted the piece up. First, my son got on the scaffold and pulled his end up while I helped by pushing. Earlier we had thrown a strong rope over the ridge and let the end trail down to the ground on the other side. We tied the rope around the middle of the cap, but if you want to get fancy, fasten it at two places to form an upside-down "Y" arrangement. I then ran around the other side of the barn (I'm no fool) and pulled on the rope while my son held the monster in place on the ladder. (NOTE: The ladder, of course, is placed in the middle of the length of the barn.) I pulled and he pushed as he walked up the ladder. The weight was not so great that he couldn't push it the last few inches to roll it onto the ridge. If I had pulled too hard or he had pushed too much, it would have been on my side of the barn pretty damn quick. With his hammer he whacked at one end until it was in the right position on the ends. As the boss on the ground, I could give him these directions because I could see the whole thing. He then reached into his apron of trusty nails and started fastening it through the shakes and into the one-by-fours all along the length of the ridge.

Now you have reached the pinnacle and experience a great surge of feeling: it's hard to tell whether it's pride or relief! A split shake roof is really something to see. It looks like a controlled textile design that fits together, and yet no two shakes are alike. I can get carried away with a sight like this. A nice old custom is to put a tree branch on top of the peak. I just said "Thank God" under my breath.

If you want a chimney pipe for a coal or wood stove, see your local stove dealer and get enough insulation around that pipe where it comes through the roof. Get good quality stuff, and give the chimney pipe the proper height to carry the sparks away and not onto the roof. (Information about how far away from the wall you should place the stove and what kind of metal or stone barriers are available to protect the wood-

Inside construction of the window frames and the siding going up. Lapped floor was also in progress.

en wall from the heat is available from several sources, including your stove dealer.) As you can see in the picture, I now have a barn with skeleton sides and a complete roof. I love it—it looks so structural and free. Now I can lay the floor without worrying about its getting wet and then go on to the siding.

FLOORS

It is very tempting to reward yourself for all the work you have been doing by slapping down those big fat floor boards in a hurry. Listen to a word of warning before you do though. Wood shrinks on the width, especially green timber wood. Unless you are going to lay a double floor, which you should do for a heated building, you may be surprised when grass starts sprouting through your one-layer floor. And even if grass doesn't grow well under a building, you will surely get a lot of air coming up through those unattractive black gaps appearing between the floor boards. I decided to make my 1" x 12" flooring boards with lap joints. If you have trundled your table saw out to the site by now, you can run the boards through the dado blades. I found this impractical because I have a heavy table saw. Use a router here instead—it will do a very nice job. I used a gouge bit that made a slot about 5⁄16″ wide and set the depth at ½″. The little porch at the front of the building with its 2 one-by-twelves made a great workbench for this kind of activity. I would C-clamp the ends and the middle, using a very straight piece of kiln-dried ¾″ x 8″ x 12′ lumber as my straight edge to guide the router. You have to figure how far from the edge this piece should be in order to position the router blade right along the edge. Again I used a trick from my roofing experience: I made myself another little stick the right length and very easily positioned my ¾″ x 8″ board the right distance and clamped it in two places.

The floor boards have to be fitted to go around the studs and floor angle braces (the first and last board). Then they need to be nailed with three nails across each floor beam. I used 2½″ galvanized common nails. If a board is warped or cupped and you want to force it down, use a long nail in the middle. Always face the curve of the

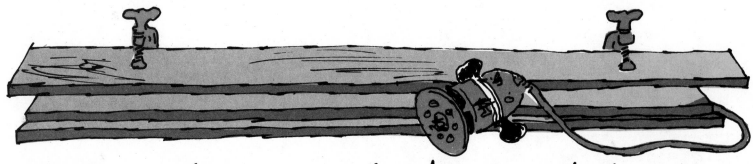

Cutting method to make lap boards with the router.

the lap joint.

Floor boards alternately lapped.

annual tree rings so the curves go down or against the beam you are nailing it to. I also staggered the boards as I did the roof slats. This lends a stronger building and bracing effect. The lap board too contributes to making a more effective brace and is stronger structurally. Now's the time to strut back and forth on your floor—it will give you a marvelous feeling, and you deserve it. Take pleasure in the beauty of this inner structure while you can, because soon it must be covered up with siding.

WINDOWS

I purchased the windows early on so I could make the rough openings the right size. I don't care much for the double-hung windows used in houses, preferring a single flat surface with small lights (windowpanes). These windows can be either nailed in a fixed position or put on hinges. I hinged each of mine on the side with a pair of hinges and added a long hook and eye to hold it open and two sets of short hooks and eyes to hold it closed. If you follow my drawing for window placement, you can put the horizontal and vertical two-by-fours in place, allowing a clearance for the windows. You will also want to decide on the style of casing you want around the windows.

When it came to the window trim, I wanted to follow the generally accepted rule requiring the trim to be 2″ thick, but I also wanted to create a different visual effect. The facing boards were one-by-fours, so I cut some 1″ x 1″ strips and lined the top of the outside edge of the one-by-fours, as my drawing indicates. I then had a more delicate window frame, yet maintained my 2″ thickness at the outer edges. Remember to leave ¼″ extra space all around the window if you want to install hinges and let the windows swing open or closed. That sounds rough—but it works because it's enough to forgive a not-quite-square installation but not so off that cracks can't be covered gracefully by trim.

A field mouse's eye view of the corner molding and how it protrudes out from the siding.

Before you put the siding on there are two small jobs you really should do. The first involves the "flashing," which is just a shaped metal covering whose job is to be a water barrier that goes over the windows and doors. The picture shows how to install it. Flashing comes in aluminum, which I used, galvanized steel, and copper. The latter two are quite expensive. I feel the aluminum is OK, especially if the exposed part is painted. I bought 36″ wide flashing and cut it down the middle to give me 18″ strips. I nailed these to the studs and then brought them down over the tops of the window and door moldings, but first I had to shape the flashing strips to fit. Using the edge of a bench, I shaped the edge into something that looks like a staircase seen from the side. I shaped it by pressing the metal over the workbench edge and then hammering it to make the crease. You will probably develop your own method for doing this. Short galvanized roofing nails are best.

The second job to do before the siding goes on is to be sure the corners of the structure have raised moldings (this includes the doors and window frames) of the proper thickness, so that your siding has something to rest flush against. I like the thickness of these pieces of wood to protrude a little beyond the siding (about ½″). If the moldings end up flush with the clapboard siding, the effect will be unpleasing to the eye. My clapboard was ¾″ thick, so at the lap (where one board overlaps the next one) the thickness was 1½″. This meant that the molding had to be 2″

The siding goes up in earnest and my beautiful skeleton is disappearing forever.

thick. I managed this at the doors and corners of the building using two-by-fours lapped at the corners. Since there are no two-by-threes in timber lumber, I cut 1″ off one two-by-four and then lapped them so that I ended up with a 4″ x 5″ corner 2″ thick.

Basic detail of 2″ x 4″ framing of window between studs.

Metal flashing goes over windows and doors.

Window molding detail.

Side of barn showing "flitch" sawn siding with all its irregularities and the window molding completed.

SIDING

When the whole skeleton of the building was complete, I had a hard time going on because I was so taken with the beauty of the bones. Somehow it's a free soul at this point as it joins with trees and sky. I framed out the front door first. Take a close look at the photographs and drawings if you haven't done it before. You'll notice that the frame is doubled in many places for strength; this frame will have to hold two heavy doors later on. You must make a decision at this point about the type of siding you want. I like the horizontal "flitch" cut planking, but you might like vertical siding instead. If you choose the latter, you really need some horizontal 2″ x 4″ braces between the studs so that you have more than just the top and bottom to nail to.

The siding I chose was the "flitch" cut type, where the sawmill saws through the log, leaving the tree's natural shape on the outer edges of the board. You should strip the bark off within a few days so that the small worms that live in the bark do not get a chance to burrow on into the board itself. (This is most important in the hot summer months.) The bark comes off rather easily—I pulled mine off with my hands. Most of my boards were about 7″ to 8″ wide. I overlapped them approximately 1″ more or less because they were irregular. You have to apply these flitches artistically. The beauty of them lies in their wavy irregularity and knots, so put one nail in the middle and step back to see if you like the lap. Be sure to stagger your butts, and always begin and end the boards in the middle of a stud for nailing. The first board should have a straight edge on

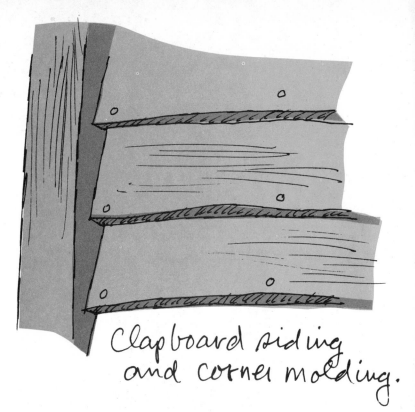

Clapboard siding and corner molding.

2″x3″ and 2″x4″ for finished corner molding.

the bottom so it lies on top of the base two-by-six cleanly.

The other type of siding you can use is the more common house siding, consisting of smooth, straight boards. Flitch siding is more reminiscent of our early beginnings in this country, and it definitely looks more primitive than the smooth boards. You can have your straight boards cut from timber ¾″ thick; if you like, you can also have it planed on one side to ⅝″ or ½″ thick. Or you can go to more expense and buy either kiln-dried lumber from a regular yard in the ½″ thickness or cedar tapered siding. The width of the exposed plank really depends on your own taste, but whatever the width, you must be careful to use a measuring stick and level to keep these planks parallel or the result will look awful. Lightly nail three or four boards up at the distance apart you think you would like and step back to see the result before you commit yourself to the whole barn. I used two nails through the board at each stud—one at the top and one at the bottom. At the corners, be sure to have enough stud wood to nail into, or put a back-up piece on first. Look at the corner illustration to get this right. Start the planking at the bottom and work your way north to the top of the plate. For ¾″ clapboards, use 2½″ galvanized common nails. When all the clapboards are nailed up, get some tubes of caulking and a caulking gun so you can seal all the seams at the corners, windows, and doors. This keeps water from entering the seams. The siding job goes pretty fast, and the boards are light enough so that you can manage them singlehandedly on a ladder. For each board, drive the first nail in the middle of the board to hold it in position so you can adjust it.

DOORS

The unique thing about the doors is the hinges. I hesitate to claim that they are original, but I came up with the idea on my own. Using metal bothered me because the hinges would rust and also they would be less than authentic. So why not wood? I thought. My pencil and paper came up with a heavy-duty hinge that enabled the doors to swing free and added greatly to the rustic appearance of the barn. The pictures and drawings show best how the idea works out. The hinges are heavily greased, and I used hardwood pegs for the pintles. (If you go with my idea for the hinges, when it's time to regrease the pintles all you need to do is take a crowbar, insert it under the door to raise it about 1″, and slide in a wooden wedge. This lifts the door easily and leaves the area to be greased open.) The hinge itself is of oak. The diagonal design of the front doors was made from the leftover ¾″ siding. I trimmed both sides straight and then lapped them with the dado on the table saw. (You can also use a router.) One-by-fours on both sides of this around the edge gave a sandwich effect and held it all together.

Some fancy measuring and cutting is necessary when you build these doors, especially with that 45-degree diagonal cut on the upper corners. You need two people to pick up each door, but the great thing in hanging these doors is that you put them in position with a ½″ strip of wood under them to give enough clearance; then you put the hinges on at that point. Slip the dowel end (pintle) of the hinge up through the hole of the door strap and fasten it to the building. I used ¾″ oak pegs for the pins and put 5″ nails and two 5″ lag screws to hold these fixtures on. A 1″ peg might be even better insurance. Grease the pins before you secure them to the building. Released from the ½″ strips under the doors, they will swing free and open completely to the side of the barn. I used two-by-fours resting on steel barn-door braces to fasten the doors shut on the inside. Three of these units should suffice. Put some one-by-twos around the inside of the door for stop molding to keep the breeze out and the door from going inward (the molding also covers any miscalculations of the door fit).

I had also considered using a vertical panel-type door instead of the diagonal. Both are pictured in the drawing. The trim that holds the paneling together (one-by-fours) should be nailed from both sides with 2½″ common galvanized nails (lots of them) or machine bolted right through with washers and nuts. This is the way the wooden strap hinges are attached to the door. I had to use several different lengths because the hinge tapers in width. (NOTE: In constructing the doors, I laid the diagonal pieces on the barn floor first and cut them to the correct overall shape. Then I put the one-by-fours on all around, first on one side, then on the other. Don't cheat and put them on only one side, as the door will surely fall apart.) The back door is made of lapped boards again but is simply braced

Note brown caulking along edges of siding.

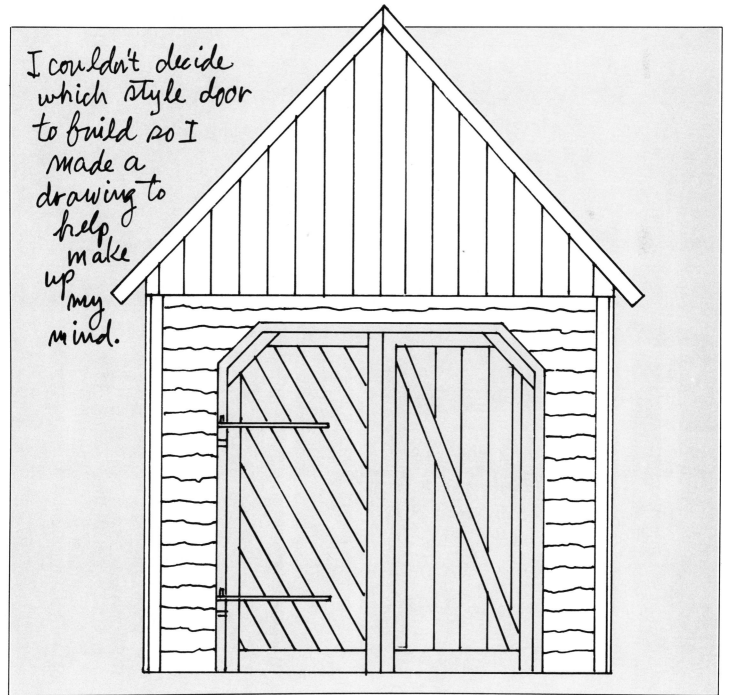

I couldn't decide which style door to build so I made a drawing to help make up my mind.

This shot really shows how the hinge is constructed, the greased peg and the bolts going through the door.

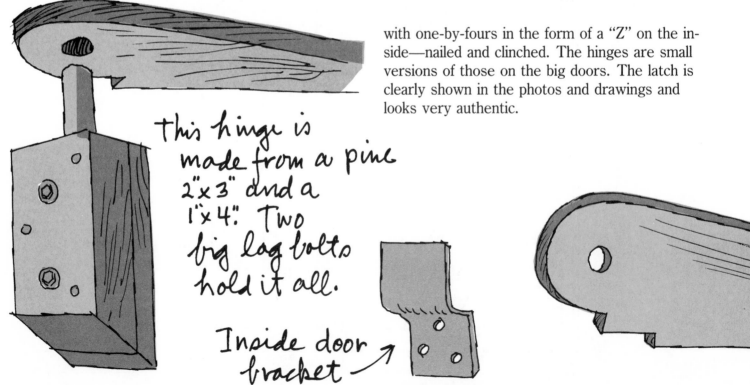

with one-by-fours in the form of a "Z" on the inside—nailed and clinched. The hinges are small versions of those on the big doors. The latch is clearly shown in the photos and drawings and looks very authentic.

this hinge is made from a pine 2"x3" and a 1"x4". Two big lag bolts hold it all.

Inside door bracket →

STAIN

I approached the staining of the barn with great trepidation because I had fallen in love with the natural color of the wood. I knew this affair would not last though, and that eventually the wood would turn blackish and uneven, finally becoming gray all over and having no protection. My wife Marcia and I liked the idea of a red barn (very traditional), so I tried to buy red stain. But the available stains looked more clay- or brown-

Oak hinge

colored than red and seemed overpriced as well. After some research and a lively conversation with a paint store owner about what stain really was, I came up with this formula: to one gallon of your choice of exterior paint, add one pint of boiled linseed oil and about one quart of paint thinner. This yields a very good paint stain which still retains some of the body of the paint. If you want, you can add a little more thinner, but the stain will shortly get very transparent and not offer very good coverage or protection. I found a very substantial barn and roof paint. The color is beautiful and has weathered two winters satisfactorily. Use a big inexpensive bristle brush, as the rough wood surface needs to have the stain brushed in. Use a small brush for the windows. DO NOT buy one of those nylon shiny brushes; they are useless. The stain will seep back over your hands and you'll need to do a lot of wiping, so keep a supply of rags handy. Wrapping a rag around the metal band of the brush helps keep this problem in check. Marcia and I did the whole barn in a weekend, even taking time out for another trip to the paint store for more supplies. I tried using a roller on an extension stick, but it had no effect whatsoever on the rough boards.

A word about why I used paint stain instead of paint. Wood is a living, moving organism, and even after it is cut and dry it still moves, stretches, and contracts. Paint does not. Stain preserves and fills in the wood pores, whereas paint usually sticks to the surface. For this reason, most wood and shake buildings should be stained, not painted. Stain, being a more fluid substance, will cir-

Back door and its early-type latch and hinges.

culate in the fibers of the wood and not form an outer crust that will flake off and peel.

Every couple of years, roll yourself under the edge of the building and treat the wood resting on the piers to a bath of green Cuprinol to keep the termites from being interested. The shakes, remember, have already been bathed in their brown stain (if you were smart).

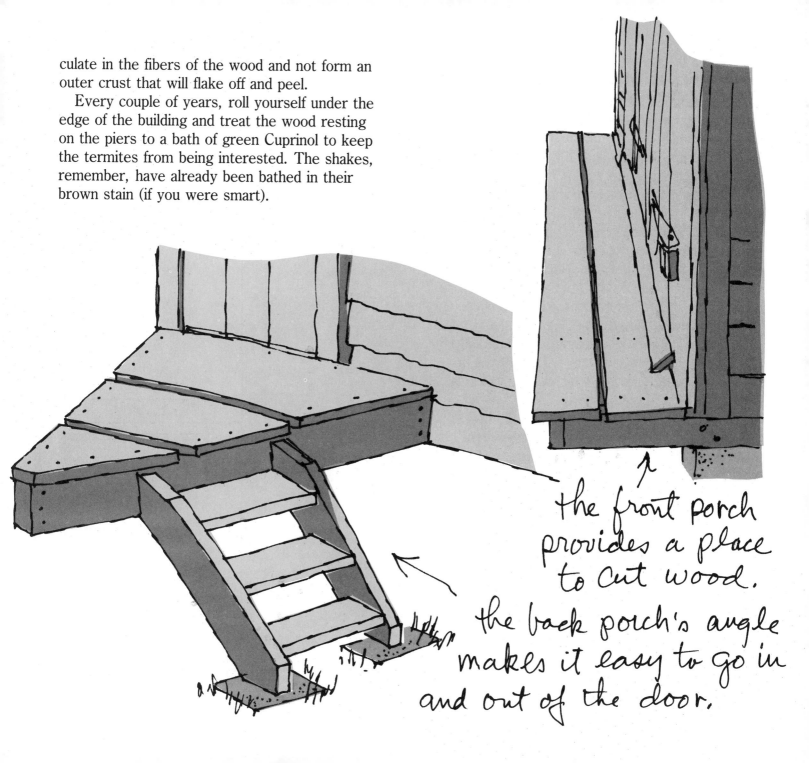

the front porch provides a place to cut wood.

the back porch's angle makes it easy to go in and out of the door.

THE GREENHOUSE

The construction plan for this part is basically self-explanatory. The four floor beams are tied into the regular floor beams with nails and some bolts. The floor beams should be long enough so that as much hangs out as ties into the barn beams. You could also extend the original floor beams out to the right length at the beginning. My greenhouse was an afterthought. The beams and sill boards need to be notched a little where they cross one another. This enables the greenhouse to "float" above the ground. This design is truly experimental on my part, and I do not know how effective it will be. The sun should warm the inside table surface for starting seedlings, and I have lined the floor and walls with insulation to retain the heat. The side windows are plate glass as they are fairly large, and the roof glass is regular window glass. All the glass must be heavily caulked because leaking is hard to avoid, especially from the glass roof section. You need strips of wood or molding behind the glass as well as in front. Put a bead of caulking along the edge of the glass before you nail the outside molding against it.

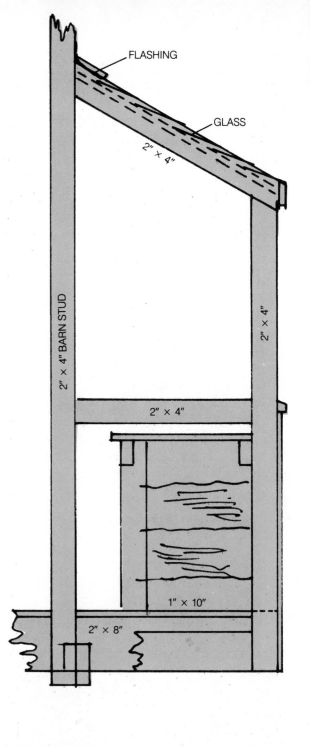

TIMBER SHOPPING LIST FOR GREENHOUSE

SILL
One 2″ × 6″ × 8′

STUDS
Seven 2″ × 4″ × 14′
One 2″ × 4″ × 8′

MOLDINGS
Miscellaneous lengths of 1″ × 1″

FACING BOARDS
Two 1″ × 4″ × 8′

FLOORS
Four main beams from barn floor 2″ × 8″ × 16′
One beam extension by lapping 2″ × 8″ × 6′
Continue 1″ × 10″ flooring from inside the barn.

SIDING
Four ¾″ × 8″ × 12′ (Flitch sawn boards)

TABLE
Three 2″ × 4″ × 8′
Short pieces 1″ × 10″ or 1″ × 12″ (table top)
Ungrouted tiles (table top)

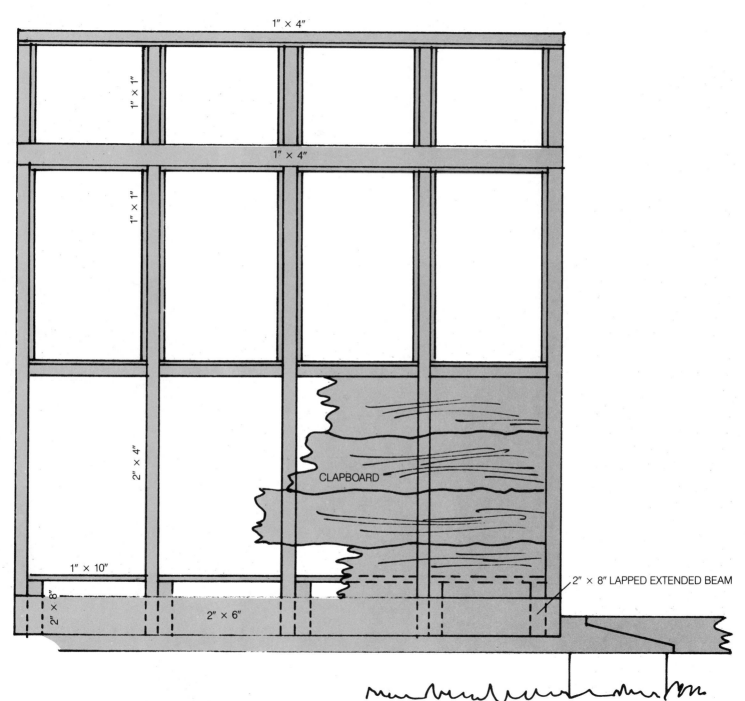

1" × 4"

1" × 1"

1" × 4"

1" × 1"

2" × 4"

CLAPBOARD

1" × 10"

2" × 8"

2" × 6"

2" × 8" LAPPED EXTENDED BEAM

81

SALTBOX SHED

Pier and its anchor bolt.

Bathing the hidden sill with the green Cuprinol preservative.

This shed turned out to be a lot bigger than it looked on paper. I am always amazed at how actual space seems so much larger than flat-dimensional views on paper. The high peaked ceiling is 11′ off the floor while the width and length dimensions are about 9½′ x 12′. I am trying to think of different ways to use that space up in the peak. It would be wonderful to hang herbs and dry things from.

Our land was farmland for many years, so we have a mixture of hardwood groves, open fields, rolling hills, and too many vines and sumacs that run wild. We have done a lot of silva-culture (tree control) by cutting back the sumac and vines to free some beautiful oak and hardwood trees that were being choked. Having done that, we created the need constantly to mow the fields to keep these weed trees from taking over. So we invested in an important new tool—a garden tractor to keep our beautiful fields from disappearing in a tangle of fast-growing vines and underbrush. We cleared one area and found that it had a natural basin for a pond with a stream flowing alongside it. I hope to do a chapter in another book on my experience in building this pond. My God, but this is a long-winded way of telling you why I designed and built this shed.

I wanted the shed to house my garden tractor, mowers, spreaders, gasoline cans, and garden tools which were hanging all over the place. The barn is being completely used as a place to build a replica of a double-ender sailboat (in wood, what else?) used by New Englanders for fishing in the 1800s. I built this shed a little more simply than I did the barn. Placed at right angles, the two structures enhance each other in a very pleasing way. Also, the shed is being stained a gray color because I think I have gone as far as I can go with red buildings.

Early in March I noticed, after turning over a few shovelfuls of dirt, that the ground was pretty soft. Apparently the frost line did not penetrate

very deeply that winter. I dug the holes for the four piers down to about 36″ before I hit shale, which is part of our terrain. I used 8″ x 8″ x 16″ cement blocks for the piers, filling them up with gravel and cement to the desired height. I used the straight two-by-four and level to find the correct height for each pier because the land sloped a bit. It was necessary to keep in mind that the tractor needed the low side of the building for the entrance. As I positioned and cemented each pier, I put an anchor bolt in to accept the wooden sill. Figure how the sill timbers will lie before you put in the anchor bolts, because a wrong position will give you monumental headaches. Even though this is a small building, still use the stake and string method of laying out the rectangle to get those piers square with each other.

SILL

After the piers have hardened, I install my favorite L-shaped sill and we are off and running. The L-shaped sill allows me to notch the two-by-four studs for extra strength. The floor joists are two-by-sixes, which are sufficient to handle the weight of the tractor and more. The joists are about 18″ apart. Don't forget that if you use full-dimension timber from the mill instead of kiln-dried lumber, you can open your spaces between studs and beams a bit. But if you use the kiln-dried lumber, increase your number of beams or

2 x 4

4 x 4

Combination of 4"x4" and 2"x4" that make up the corner post.

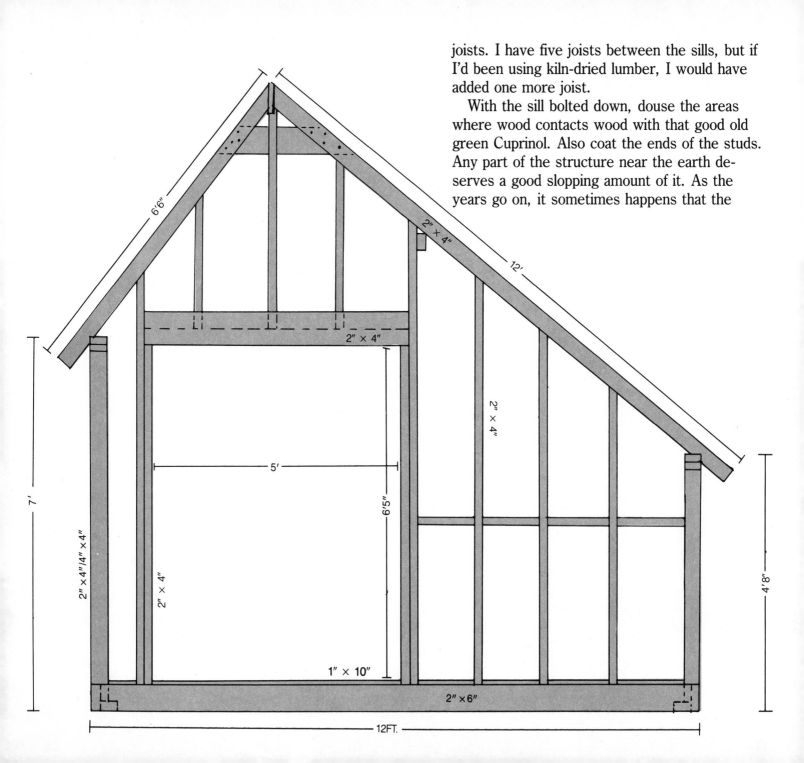

joists. I have five joists between the sills, but if I'd been using kiln-dried lumber, I would have added one more joist.

With the sill bolted down, douse the areas where wood contacts wood with that good old green Cuprinol. Also coat the ends of the studs. Any part of the structure near the earth deserves a good slopping amount of it. As the years go on, it sometimes happens that the weeds, grass, and soil creep closer to low areas in the sill and siding. This sill at the upper side of the shed is a bit closer to the earth than I like, but I intend to keep the weeds down. And every year in late spring I'll get down and make sure termites haven't made mud tunnels, adding more Cuprinol to the wood at these corners so that if they do it they will get a poisonous cocktail for their trouble.

STUDS & RAFTERS

The studs are toenailed to the sill. First put four-by-fours at the corners and then add a two-by-four to fill it out flush with the sill. I did this to accommodate the type of siding I used later. After the corners are up, put 2 two-by-fours on top of each other to form the plate on both sides of the structure. (I like sturdy buildings.) Then put in the in-between studs and toenail them at

The saltbox shed called for a different setting up approach than the barn because of the different roof angles.

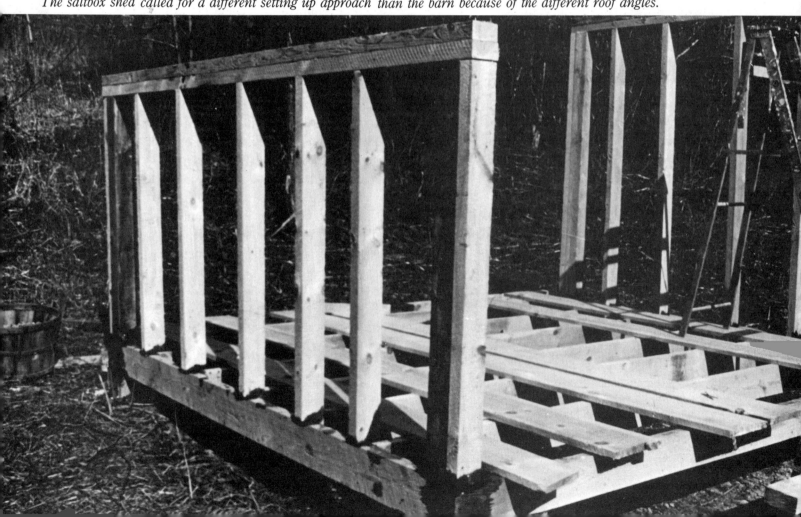

top and bottom. At this point you must leave the stud work and jump to the roof rafters. You cannot do the studs on the two sides of the building where the peak is formed simply because there is nowhere to nail the studs without a rafter.

These rafters are two-by-fours and meet at a peak board which is a one-by-eight. Here is where the job gets tricky. You must get the two roof angles off the plan, using your bevel tool. The angles are different because the two slopes of this roof are different. You figure the length of the rafter, the angle of cuts at the peak, and where the bird's-eye notch should go. Precut all this and then assemble. The assembling goes like this. The two-by-fours are light enough so that you (with the help of someone on a ladder holding it) can toenail one rafter and then the opposite rafter at the bird's-eye junction with the plate. They can be left leaning against each other at the peak temporarily as you go and do the opposite set of rafters on the other end of the roof.

Once you have these two leaning against each other, you are ready to connect them with the 1″ x 8″ peak board. Standing on an A-ladder with someone holding it steady, put one 3½″ common nail through one rafter into the peak board. Do not nail this all the way home yet—just enough to hold. Rest the peak board on your ladder at an angle because you are putting only one nail in for now. A better method is to hold the peak board up from the ground with a long one-by-four while a friend on the ladder nails. This is really the only part of this job where a helper is truly a big help. You should now be able to move the ladder over to the other side of the building where you can join the rafters to the other end of the peak board. One through nail and one toenail should hold each rafter to the peak board strongly; add another toenail on the other side if you wish. Then simply add each rafter until you finish. I used the first set of rafters as patterns for all the others. That assured me that all the rafters would be uniform in cut and notch positions. I was then able to go back and complete the wall studs as you see them in the plan and pictures.

The tops of the studs were notched into the end rafters for greater strength. This called for cutting 2″ notches and two angles on each piece and is not as hard as it seems. First stand the stud up alongside the rafter and pencil mark the angle using a level vertically to make sure the stud is straight. Make a mark on the new stud above and below the rafter edge. Cut the 2″ with-the-grain cut with the circular power saw and use the hand saw to cut the cross-grain cuts. This can be toenailed from several angles very nicely. It is a great strength joint and very pleasant to see installed.

I happened to have a storm window with many lights (windowpanes) in it that was perfectly proportioned for the wall, so I framed out the 2″ x 4″ studs to hold the window (allow ¼″ all around for easy fitting). Order your window early so that you can be sure to get a stock size; otherwise you may find yourself paying a lot extra for a special size. Just a word to the wise: don't put the window in place until after you have put the

siding on the frame as—guess what?—the glass will sometimes shatter with all that hammering. I used 1″ x 1″ strips of wood for the inside stops to hold the window in place. If you want a moving window, more work is involved. A 2″ window sill beveled to shed water is a good alternative.

After the vertical studs are in place, you will need some horizontal two-by-fours to give you another place to nail the vertical siding if you are going to elect this type of siding. With clapboard siding you don't really need them. This shed would also look marvelous in wood shakes to give more of a Cape Cod look. Use the smooth, shorter wood shakes that are normally used for siding work. The roof shakes shown on this shed are hand-split, and I think they look better on this particular roof than the smooth ones would. They are about 22″ long and the smooth ones about 16″. The roof as well as the sides can be done with the smooth shakes and look best with a gray finish for a weathered look.

THE ROOF

Now you are ready to put the slats on the roof prior to shingling. (The reason I like to finish the roof before putting on the siding is so that I can lay the floor out of the weather. It's easier to lay the floor with the siding off, too, because you can squeeze in and out between the studs and fit the floor boards with more freedom of movement.) The roof slats are made of one-by-fours laid 9″ on center to give the shakes a 9″ amount of open

The studs and roof beams without a cross brace waiting for the vertical studs to be inserted.

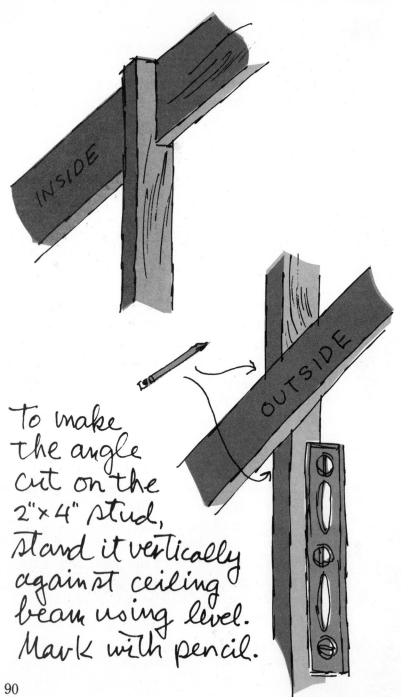

To make the angle cut on the 2" x 4" stud, stand it vertically against ceiling beam using level. Mark with pencil.

surface per course (row). You can make it 7" or 8" if you like, but that will require more shakes, whereas more than 9" looks too wide to my eye. But whatever the amount of surface showing, the 1" x 4" slats should be the same amount on center.

I neglected to mention the facing boards that cover the roof slat ends. I made these from one-by-fours, and they can be the dickens to nail up. I would take a bar clamp and clamp it to a stud up near the peak to rest one end of the one-by-four on and then nail the bottom end. The big problem is locating it so it centers perfectly on the center peak board. A Jorgenson clamp near the peak allows you to position the end at the center of the peak. Add another clamp and nail. Leave enough timber to cut the lower end after it's nailed in place. Do this before shingling be-cause the shakes overlap this board. The one trouble with this design is that with two different angles meeting at the ridge, one board will ap-pear wider than the other. Fit a ¾" triangular piece of wood up there and bring the two facing boards up to the base of the triangle to cover the butt. The easiest way is to run one facing board all the way and butt the other one to it at what-ever angle is called for, as the drawing indicates.

SHAKES

Use smooth shakes for the first course, then over that lay the first real row of hand-split shakes and proceed to the top ridge. The first row is the only double tier. Use a measuring stick 9" long to check the rows as you go instead of a batten across the roof. In this manner you can lay the roof from right to left or left to right

Window framing all done and roof slats without their facing boards.

from top to bottom without worrying whether the rows will look straight when finished. The split shakes look better if they seem to jog along rather then giving too mechanical an effect. If you are roofing or siding with smooth shakes, it is better to use a nail-on batten stick to butt them against because they look smart lined up evenly.

Most of the shakes can be put on by walking up the slats until you get near the end of the roof; then you will need the extension ladder.

There's a trick to using the ladder for roofing, and the photo showing the ladder illustrates it pretty well. You angle the ladder against the roof at about the same angle as the roof slope

91

Sitting on the peak of the saltbox while putting on the shakes.

TIMBER SHOPPING LIST FOR SHED

This list should cover just about everything you will need for the Saltbox Shed give or take a board or two or three.

SILL AND FLOOR BEAMS
Seven 2″ × 6″ × 10′ (floor beams)
Two 2″ × 6″ × 14′ (2 sills for long sides).
Two 2″ × 4″ × 14′ (inside sill)
Two 2″ × 4″ × 10′ (inside sill)

FLOORS
Five 1″ × 10″ × 10′

STUDS
One 4″ × 4″ × 10′
One 4″ × 4″ × 14′
Four 2″ × 4″ × 10′
Four 2″ × 4″ × 14′
Ten 2″ × 4″ × 12′

PLATE
Four 2″ × 4″ × 10′
One 2″ × 4″ × 10′ (cross beam)

ROOF
Seven 2″ × 4″ × 10′
Seven 2″ × 4″ × 12′
One 1″ × 8″ × 10′ (center peak)
Two 1″ × 4″ × 12′ (face boards)
Two 1″ × 4″ × 8′ (face boards)
Twenty-six 1″ × 4″ × 10′ (slats)

CAP
Two 1″ × 6″ × 10′

DOOR
Five 1″ × 6″ × 14′
Two 1″ × 6″ × 12′ (braces)

SIDING
Fourteen 1″ × 8″ × 8′ (short end of shed)
Seven 1″ × 8″ × 14′ (higher end of shed)

Window side:
Seven 1″ × 8″ × 8′
Seven 1″ × 8″ × 10′
Four 1″ × 8″ × 12′

Door side:
Five 1″ × 8″ × 8′
Five 1″ × 8″ × 10′
Two 1″ × 8″ × 12′
One 1″ × 8″ × 14′

Eighty 1″ × 1½″ battens varying lengths

ROOF SHAKES
Two SQUARES plus two BUNDLES
One BUNDLE smooth shakes

FOUNDATION
Sixteen to twenty 8″ × 8″ × 16″ cement blocks
Three 80 lb. bags concrete mix.
Sometimes you will need some 4″ × 8″ × 16″ cement blocks to make the piers the right height.

The tractor tries out its new home while there is still a view.

Two ways to join the peak facing boards when roof has two angles.

but slightly upward so you don't press down on the shakes that are under the ladder. BUT, the most important thing to do is to put two sturdy cement blocks against each leg of the ladder; this seems to be a foolproof way of keeping the thing in place, especially on sloping land. Otherwise it can slide on you as your weight gets higher up the ladder. Do not use the rubber feet that aluminum ladders come with—they slide on dirt. Instead, tip the feet upward so they act as a brake and dig into the soil. Extend the ladder as you progress toward the ridge, and shift it along sideways as you progress from, say, right to left. With your nail apron and hammer belt holder you should be all set and confident for this kind of work. Leave anywhere from a ¼″ to ½″ distance between the shakes to allow them to expand when wet and for breathing space.

I made the ridge cap on the ground and pushed it up the ladder ahead of me. Build it with a one-by-six and a one-by-five, and remember that for any angle but 90 degrees you will need to make an angle cut on the 1″ x 5″ board. The cap can be stained before installation, both inside and out, for longer life. Be sure to measure for the ridge board from the actual roof and not from the plan, because these dimensions change as you build. Notice in the still-life photo on staining that I prestain the shakes about halfway up their length. This gets the maximum amount of protective stain on them, and it's a lot easier than staining the finished roof.

Later, when I could maneuver after the floor was laid, I decided to put cross braces up near the peak just to keep the roof rafters from spreading since there aren't any cross braces at the plate. To do this, again use your bevel tool to make the two angle cuts. My braces were about 30″ long on the bottom edge. Put one on every rafter or on every other. Thinking ahead, it's a good idea to put a two-by-four across under the rafters on the longer sloping roof about two-thirds of the way up. This board will act as a support for snow and ice that may lie on the roof and as a protection against time, which has a way of causing things to bend and sag. Nail the ends of this support to a vertical stud on both sides of the building.

More stain and more drying shakes.

Shaking technique going from right to left.

FLOORS

Since this shed is small, I was able to cut each floor board out of a 1″ x 10″ x 10′ plank. This time instead of using the router to lap each board, I took all the boards up to my shop and ran them through the dado on the table saw. In my situation I have to heft the planks up to a second-story window because my woodworking shop is upstairs, but it was well worth it because the dado is easier to use than a router for this.

The lap floor is strong, and when the wood shrinks you still see wood and not the ground underneath. I nailed three galvanized 2½″ common nails across each board into each floor joist. It will take you some time, but you must cut the two outside boards around the studs. Just lay the floor board up against the studs and mark the thickness of each stud where it intersects. Then mark off the 4″ depth of the stud and cut all the notches out. The board should fall right in there. Don't forget to make an allowance for the last board in case it has to be cut narrower before marking the notches. I love the feeling I get after the floor is laid; it's somehow so satisfying to walk on your own floor, rather like experiencing your first tree house all over again. I like the solidity and looking through the studs at the view before the siding hides it all. I do go on about this kind of thing, but part of any work should be these moments of unbridled satisfaction with your creation. This is the real personal reward, even though the moments are fleeting. I regard them

Covering boards are nailed to the sill and studs.

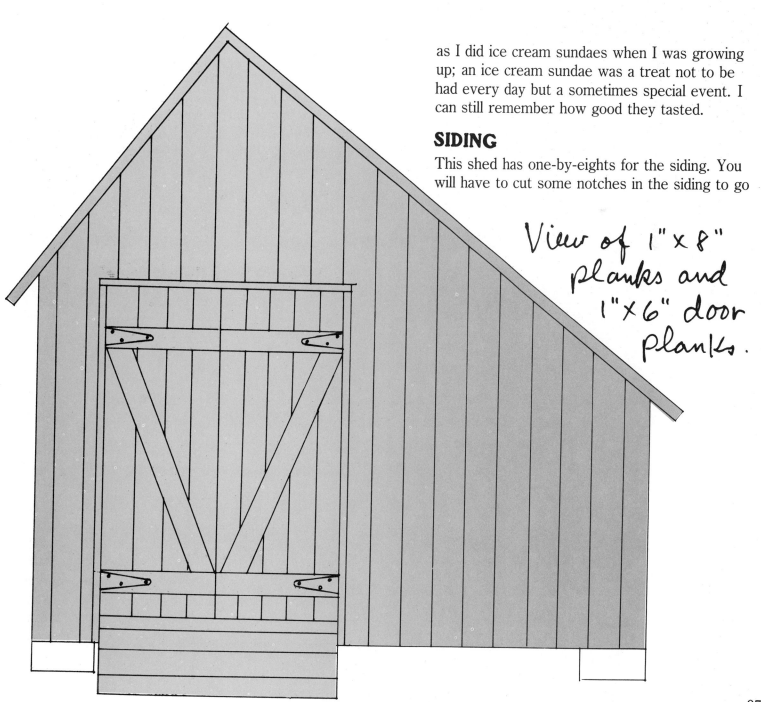

as I did ice cream sundaes when I was growing up; an ice cream sundae was a treat not to be had every day but a sometimes special event. I can still remember how good they tasted.

SIDING

This shed has one-by-eights for the siding. You will have to cut some notches in the siding to go

View of 1" x 8" planks and 1" x 6" door planks.

Two ladders for two different heights.

The 1" x 1½" battens close up the slits between the covering boards. Nail into one side only to allow for expansion and contraction of covering boards.

the rough opening of 2" x 4"s can be capped with 1" x 1½" molding to line up with battens.

around the rafters, but you can side high up the wall until you touch the ceiling slats. On the sides of the building you will have those angle cuts to make, when the bevel tool again becomes the hero of the day. Finishing off the shed with batten boards gives a whole new effect and also seals any shrinking of the 1" x 8" vertical panels.

The battens give the inside a cosier feel and keep the howling snow out. These were made of 1" stock 1½" wide. For wider siding, the width of the battens should probably be increased a little—say, up to 2". If the siding is much narrower than 8", the batten method will require too many. I used 2" galvanized roofing nails for putting the battens on. I suggest that you put the nails through the battens into one wall plank only; otherwise the expansion and contraction of the planks may split the narrow battens. After this task is completed, the only openings left in the building are those 1" by about 5" openings between the slat boards on the roof ends. I left these for ventilation, but they can be stopped up with wedges of wood. Gluing these wedges in is probably easier than nailing them.

DOORS & RAMP

I made the doors of one-by-sixes and braced them with the same, because I liked the contrast with the 1″ x 8″ siding. The 1″ x 6″ braces form two opposing "Z" designs, that together make a large "V" configuration on the outside of the doors. Clamp together the boards that make up the door with two bar clamps (Jorgenson or other); then nail from the outside in and clinch them on the inside. Remember that using the lap joint for these boards is best. When you are nailing, put a couple of boards under the door out of line with the nails, so that you are elevated and thus you can avoid nailing into the floor. Take the circular saw and trim the edges to make them all flush. Use 6″ or 8″ metal strap hinges—not the double-strap kind, but the single strap with a regular door butt plate. This way you can use them in a way they are not meant to be used.

I'll explain. These hinges are made so that both the strap and butt are supposed to lie open flat against the door and jamb. Anybody can unscrew this arrangement, and it doesn't look so hot either. Instead, turn the butt plate part of the hinge into the jamb (the space between the door jamb and the door itself) so that when the door is shut, the butt plate is completely hidden. Today, large strap hinges with a removable pin so you can reverse the butt plate are hard to find; therefore, you will have to bevel the screw holes on the wrong side, using an oversized bit in your power drill, so you can put flathead screws in. In other words, the butt plate comes with bevels

The "Z" braces facing one another making the bracing visible as well as functional.

Hardware and detail from above.

The hinges showing the carriage bolts heads.

ground into one side of it to accommodate the heads of the screws; but what I suggest doing is to use the other side of the butt plate and rebevel. Put carriage bolts that have smooth oval heads through the strap holes. The doors will swing a full 180-degree arc. I used a metal hasp for a lock; but an oak latch arrangement would look really great, so I have made a drawing of one.

The ramp for the tractor to enter the shed turned out to be a nice little addition to the architecture of the building. The door opening is roughly 5′ wide and the ramp is the same. First I buried three cement blocks of the 4″ x 8″ x 16″ type (the solid ones) in the ground approximately flush with the soil; then I shaped 4 two-by-fours as small beams for the ramp. The drawing of this explains the operation better than words do. These two-by-fours were notched to fit over a two-by-four that is nailed to the sill and then tapered to almost a point to give the proper slant to the ground. Then it's a case of planking the top with 1″ dimension boards of any width you want. The shape of the beams depends on how much slant your situation calls for from the ground up to the floor of the shed. Soak the beams and bottoms of the planks with plenty of Cuprinol. Platforms with a lot of usage can be replaced easily along the way.

The nice thing about this shed is that you can walk in and move around freely without banging your head on the ceiling; you can build a shelf along the low side and still have plenty of room to put in a lawn mower, a snowblower attachment and a spreader, etc. The shed also has room on the floor next to the tractor so that you can work on your equipment inside. I see sheds that some outlets along the highway sell, and I'm appalled at how little they give you for the money—not to mention at how these cookie-cutter constructions look. One of the reasons I'm compiling this book is to inspire people not to take the easy way out by buying some aluminum monstrosity to mar the landscape.

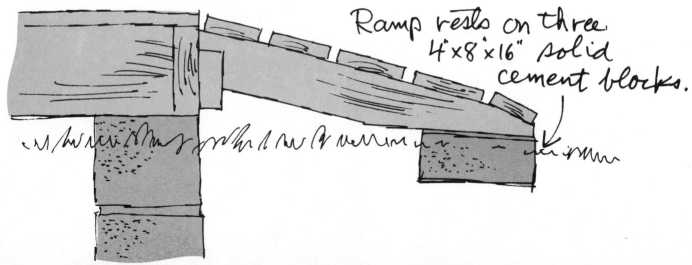

Ramp rests on three 4″x8″x16″ solid cement blocks.

Idea for rustic oak latch.

Barns and sheds emanate a feeling of beauty when
they nestle with trees and grass.

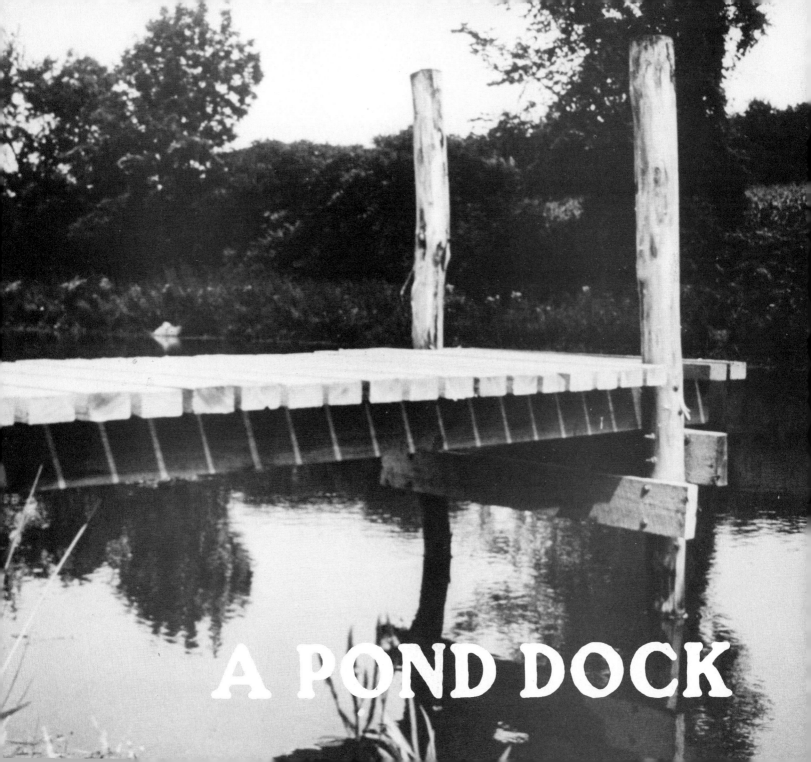

A POND DOCK

This project confounded me at first because of the length of the poles. A pile driver is not exactly a tool one has around the house or garage. It was my older son who, being an architect and having an analytical, logical mind, came up with the idea of how to get the posts driven into the pond bed.

First of all, I wanted the dock to look very rustic, as if it had been sitting in the reeds for about ten years. Locust wood posts seemed to offer the right look. You can go into the woods and look for locust trees yourself, or you can do what I did: find someone who cuts them for farmers who use them for fencing in their cows. The someone I found turned out to be a very interesting senior citizen who, in addition to delivering two locust posts about 8′ long and about 5″ thick, gave me all kinds of information about the house I live in. Locust posts come with bark, and it's a good idea to get it off before you start building so that your bolts will cinch up to the wood itself, because the bark will eventually fall off and leave a loose fit. (I didn't begin this project right away, so removing the bark was no problem for me; it was quite loose after the posts had lain a few weeks in the field.) Use a circular power saw to cut points into the end of each post to make driving them in easier. Four cuts, one on each side, should do.

Locust has always been a good choice for fencing because it is hard. If you don't have a battery power drill, you will see just how hard it is when you go to bore holes with the auger bit and hand brace. The holes for the bolts really have to be drilled at the site once you get the posts in, and ponds are never near electricity. It's kind of fun, though, standing in water up to your waist, drilling away and watching the wood chips glide down into the marine depths.

Anyhow, the really tricky aspect of this job was how to knock in the posts with a sledgehammer or, more to the point, how to get high enough to knock them in. We nailed a crosspiece (2″ x 6″ x 52″) of oak temporarily into the two posts to form an "H" at about the height we thought the crosspiece would end up. Then we dragged out the two support timbers that would attach this "H" to the land, laid them on the crosspiece, and temporarily nailed them to the posts. These were 2″ x 6″ x 14′ timbers of oak that would eventually become the real support timbers for the gangplank or deck. When using this kind of timber, take a look down its length with the flat 6″ side facing up to see which way it curves. If it's straight as a yardstick, fine, but oak usually is not quite that straight, so use the concave side facing down toward the water. This bow in the timber is an advantage because it's somewhat stronger when used this way—comparable to an arch. If the curve is very pronounced, you may want to cut the convex side to a straighter line and leave the concave curve on the bottom so that the 2″ x 4″ planks on top will lie better. I put a third timber in the center as additional support, and the whole thing feels very firm.

Once my son and I had this temporary rig in place, we threw a few 2″ x 4″ planks on top of the supports; one of us was able to stand on the planks and swing the sledgehammer, knocking

The screw eye and anchor bolt connection for real.

All three cross beams resting on their piers.

A ladder made from two 2" x 4"s and 2" x 3" for steps.

Positioning the cross beams for the dock.

the "H" right into the muck. The weight of the rig kept everything in place while the hammering was going on, but I stood in the water and steadied the poles just in case they decided to pitch over. My son pounded the poles—first one, then the other—only a few dozen times to send them biting into the pond floor. I would say we went down about 12″ to 18″. This may not seem like much, but once the 14-footers were anchored on land, the "H" took a nice firm hold. We have a lot of shale in our soil, but in sandier places you may go down farther. Two winters later the posts still stood firm in the pond's bed.

Using a level, we determined where the first crosspiece would best sit, figuring the proper level we would want for the support timbers. The two outside support timbers were to be fastened to the inside of each pole, sticking out about a foot beyond. Small bar clamps are very useful to attach the crosspiece while you drill the holes. I used two 8″ x ⅜″ galvanized machine (8-sided head) bolts for each connection with washers on both ends. Since there is no electricity near my pond, I used the old auger and brace technique. Two minor difficulties arose: first, I was standing in water up to my buns and, second, I was drilling at an uncomfortable height. I realized later that standing on a cement block in the water would have been smart. (It's amazing how hard it is to work with your arms above chest height, so try to maneuver so you can come at the work from above.)

I put two bolts into each pole through the two-by-sixes. Notice in the pictures that these two-by-sixes hang out either side. If the wood extends beyond the bolts the brace is stronger and the wood less likely to split. I then put another 2″ x 6″ x 52″ brace, lower down on the inside of the poles but still above the water line, to minimize the sideward sway that you never lose completely. This definitely stiffens it up though, and I screwed two 4½″ galvanized lag screws into each post for this purpose. (Be nice to your guests who swim by facing the head of the machine bolts outward and the tails inward, so the swimmers don't cut themselves while swimming near the dock.) I also double-bolted the two 14′ supports into the poles and rested them on the cross brace.

Back on land, in the meantime, two little trenches were dug to hold typical 8″ x 8″ x 16″ cement blocks. I used one block to support each of the three 2″ x 6″ x 14′ timbers. I cut two 24″ cold bar lengths of steel (any thickness will do) and pounded them into the soil inside each block to tie the blocks to the earth. Cold bar steel rods are sold in lumber yards. You can identify them by the raised perforations on their sides. They are ordinarily used in buildings as the skeletal structural backbone inside cement columns and cement floors—hence the term "reinforced concrete." When the rods were in place, I filled the blocks with cement. In the two outside blocks, I put anchor bolts up close to the timbers—but not before screwing a hefty screw eye into the side of each timber. This let me put the anchor bolt in the wet cement, position it through the screw eye, and hand tighten the nut and washer to anchor the oak support. Later, when the cement hardened, I tightened up the nuts. The middle timber just rests on its block and also on the brace out by the poles. Its only role is extra support for the 2″ x 4″ planks I nailed into it later.

My gangplanks were 45″ long two-by-fours of pine, precut and brought to the site. There were a couple up between the poles that only measured 35″. Note also that the 14′ support timbers were about 34″ apart, so the gangplanks hung over about 5½″ on both sides. A deck that would be flush with the supports needs to have holes predrilled for the nails; otherwise the ends will split. This kind of gangplank, if you choose to build it, would look good with two-by-sixes running the length of the dock flush with the tops of the two-by-fours.

I found it best to start on the land with the decking planks and work my way out to the poles. I tacked the first one square with the support timbers using a triangle and a 1″ piece of stock as a spacer. I tacked one nail part way into each side, but didn't drive them all the way home because I wanted to make sure that the gangplank or deck looked straight before I double-nailed each end (those galvanized 3½″ nails are tough to dislodge). I drove one nail into the middle support timber and used the piece of 1″ stock to align the ends of the boards as I went along. HINT: Wait to do all the nailing until after the cement is set, so it will set without all that vibration (a week is about right).

A final touch is to put a cap of either lead or copper on the top of each post. I cut a round piece of the metal about 1¼″ to 1½″ wider than the post face, snipped a little into the edge all around and folded them down the sides of the post, and nailed through them into the post (copper or galvanized nails will do). A cap like this prevents water from entering the end grain and freezing and splitting the post prematurely in its life.

Now that the dock is all finished, and people are sunbathing and diving off it, I am thinking about building a small deck on the land leading up to the dock for everyone to dry off on. And of course, after that, why not put up a gazebo nearby so we all can swat mosquitoes in style?

2″ x 4″ treads seemed like the right width for the proportion of the dock.

An aluminum metal cap for the locust poles keeps moisture out of the end grain. Galvanized roofing nails make good hold-downs.

TRUSS BRIDGE

While I was building this bridge, the tune from the movie "The Bridge over the River Kwai" kept coming into my head. I now appreciate what the British soldiers went through to build a bridge like that under those circumstances, only to destroy it after all that work.

This mini bridge is meant to span a small stream. Its creosote coating is finally looking natural after one year of weathering. I creosoted the timbers because of the moisture problem near water, but I'm not sure it was all that necessary. It is wise, though, to use a stain or clear Cuprinol to help preserve it. (I understand that as of January 1985 all forms of preservatives containing pentachlorophenol were supposed to be taken off the market—Cuprinol and a few others do not contain this substance. Whatever effects pentachlorophenol had on the environment, I do know it was irritating to the skin and washing up immediately after using it was a smart thing to do.) So long as the growth of grass and weeds is kept cut down to a minimum around the wood, there is no reason why it shouldn't dry out easily and thereby keep from rotting.

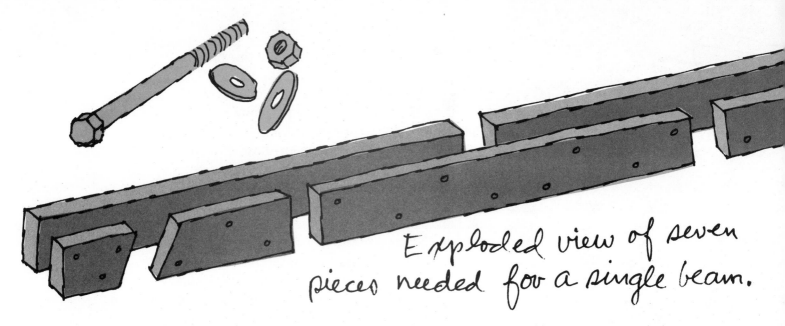

Exploded view of seven pieces needed for a single beam.

The span is about 20′ long and rests on four cement piers, two at each end. Since streams rise and fall at different times of the year, mark the spot where your stream rises to its highest level on the bank; otherwise you may be putting piers in the water instead of the shore. The height and flow seem to be the greatest in early spring, from melting snow or heavy rains. Summertime marks the dwindling point, so late summer or fall is usually the best time to do this project.

I had a very difficult time digging in this area; again the shale was tough and I had tree roots on one side. I cut some of the roots so they wouldn't heave the piers. (No, I didn't kill the big old tree—it's fine today.) Each foundation pier is sunk down about 2′ into the earth. They should extend up above the earth enough so that the wood beams will clear the dirt. Don't let a bridge

stick up too high if the terrain is fairly flat. A higher bridge looks OK when the embankment on either side is steep or a ramp is needed. The cement blocks should be filled with cement and joined with a piece of cold bar steel rodding hammered inside the blocks and down into the earth to give the pier some stability and tie it together. Anchor bolts with a threaded end should be put in each of the four piers to accept a nut and washer arrangement that is explained a little farther on.

The floor planks are pine two-by-sixes but I decided to go with oak for the main support beams. Oak is harder than pine and less likely to rot or sag as soon. Each of the two beams is made of 4 two-by-eights, each 10′ long. One 10′ length is sawed in half. The three 10′ and two 5′ lengths are bolted together in a staggered pattern as pictured on plan, so they make a 20′ long

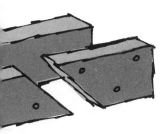

span 4″ x 8″ thick. It is pretty hard to find 2″ x 8″ x 20′ lengths of hardwood these days, so doubling up the oak two-by-eights not only yields the required length, but the strength of the staggered construction makes a stiffer bridge that can take heavier loads. Use 5″ x ½″ galvanized machine bolts, washers, and nuts to join the pieces together.

Since bridges and docks are seldom located near a source of electric power, prefabricate the pieces near the house and assemble them at the site. To assemble in one place and transport it also calls for many hands and big trucks, and who needs that for a mini bridge? If you need a span a bit longer than my 20′ one, you could still use 2″ x 8″ stock in 12′ or 14′ lengths, but beyond that you would probably need to go to 2″ x 10″ or 2″ x 12″ stock. But be warned, they are so heavy you will need help carrying the precut pieces and lifting the fabricated beams. A significantly larger bridge might also require angle braces underneath at each end, and a higher foundation would be needed at either end to accommodate those braces. I can't be sure my design for this 20′ bridge would work if your span must be much longer, so you may want to check other sources. I can recommend one book that is great to have anyhow: *Build It Better Yourself,* published by

LUMBER LIST FOR A 20′ TRUSS BRIDGE

OAK:
Eight 2″ × 8″ × 10′ (for two beams)
One 2″ × 4″ × 10′ (for two struts)
One 2″ × 4″ × 10′ (for two braces)

PINE:
Four 2″ × 6″ × 10′ (for four trusses)
One 1″ × 4″ × 8′ (for two straps)
Sixteen 2″ × 6″ × 10′ (for thirty-two treads)

HARDWARE (preferably galvanized*):
Four 12″ anchor bolts
Four heavy-duty screw eyes to fit anchor bolts
Twenty-six 5″ × ½″ through bolts for beams
Eight ⅜″ × 2½″ lag screws for straps
Four 5″ × ½″ through bolts for struts
Eight 4″ × ⅜″ through bolts for straps
Four 4″ × ⅜″ lag screws for braces
Nuts and washers for both ends of bolts and washers for heads of screws.

Rodale Press, Inc.

The truss in this design acts as a brace much like a diagonal brace in a barn construction. It is a kind of reverse or upside-down brace, if you will. The theory behind a truss is this: as the main beams sag down in the middle, when weight is put on them their ends must go up. The braces then are forced to press against each other and act as a wedge to keep the ends down, hence the middle can't sag.

The base of each of the four 2″ x 6″ angle braces is also spliced flush into the main beam construction so that it is further wedged for strength. I can jump up and down on this bridge and feel only a small tremor of movement.

MAIN BEAMS

Each of the two main beams has an inventory of five pieces: namely, 3 ten-footers and 2 five-footers. The illustration shows how to cut the angles in the five-footers to accept the 2″ x 6″ braces. At the prefabrication site I figured and cut these pieces, then clamped all the pieces together on a flat surface and predrilled them for the bolts. I marked the wood with a crayon line across each seam (or use a numbering system) so I could join them correctly at the site, where I bolted the whole thing together. The weight of each beam with its truss or brace attached is just light enough to be maneuvered by two strong men. Once it is assembled, each one must be dragged across the stream so each end rests on a pier. We dragged it down the bank and let one end rest on the near pier; then, standing in the

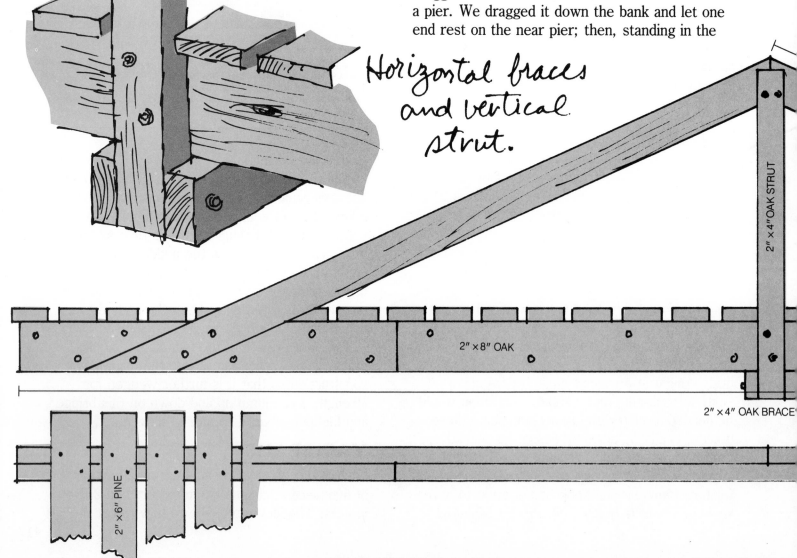

Horizontal braces and vertical strut.

2″ × 4″ OAK STRUT

2″ × 8″ OAK

2″ × 4″ OAK BRACE

2″ × 6″ PINE

stream, we slowly pulled that end across the water to the opposite pier.

Once the beams were in place, I caulked all the seams between the laminations of the 2″ x 8″ lengths. I needed only to apply the caulking across the top of the seam to keep most of the rainwater from entering the seam. (By the way, did you know that fresh water in enclosed areas tends to rot wood more than salt water? Actually, salt water sloshing around in the bilge of a boat helps preserve the wood. Most boats develop rot from condensation, which is of course

Strap connection at the intersection of the two trusses and vertical strut.

10 FT.

2″ x 6″ PINE TRUSS

20 FT.

fresh water. Any areas kept moist with fresh water and that harbor little light and air tend to create a temperature that encourages the spores of the fungus that rots wood to develop.)

TRUSSES

Here's how to figure and cut the angles at the ends of the 10′ pine 2″ x 6″ trusses where they meet in an A-shape at the top, and at the other ends where they are spliced into the ends of the main 4″ x 8″ beams. In a level area, clamp together the 3 ten-footers and 2 five-footers you cut for each main beam so each becomes a single 4″ x 8″ beam 20′ long. To measure the angles at the ends of the two 10′ trusses, where they meet at a point like an A in the middle of the bridge, use a bevel to take that angle off of the plan in my illustration. Transfer this angle to one end of each pair of 10′ trusses and make the four cuts. Clamp together the cut ends of the two trusses, butting the two angles you have just cut. Lay them flat on the ground, exactly centered over the clamped 4″ x 8″ beams with their uncut ends laid over the ends of the beams. When everything measures equally on both sides of the braces, mark the angles the trusses make where they lie across the 5′ pieces at the ends of the beams. Unclamp the marked 5′ lengths from the beams and cut them so the truss will lie flush. Trim flush any excess overhang at the ends of the braces. Be sure of your measurements before cutting so that the trusses will not end up lopsided. I think the joint looks much better with the splice in the 5′ piece on the *outside* face of the laminated beam. If it were spliced into the 10′ length on the *inside*, the walkway would be narrower as well.

Once I had both beams assembled with the trusses bolted to them, I attached the anchor bolts to the ends of the beams. The photo shows how this was done, but I will also explain it. Once the beams are resting on their piers, it is easy to see where the anchor bolts stand against the wood. Since the bolts were sticking up about 2″ I screwed a heavy-duty screw eye into the beam about 1″ below the top of the bolt. I drilled a pilot hole for the screw eye and turned it all the way to the neck (a heavy screwdriver is a great lever for screwing these babies home) and then I hefted the beam up and over the anchor bolt so the eye slipped over it. All that remained was to slip on a washer and tighten down the nut. Let the cement set for a week before tightening the anchor nuts. I painted all the bolts and nuts with a metal paint to prevent them from rusting. This should be done again periodically. I do it by getting down on my knees and firing a spray-can metal paint underneath the span.

STRAPS, STRUTS, BRACES

Once the beams and truss pieces are in place the heavy lifting is done, but the heavy twisting work begins. It is hard because there are four oak two-by-fours plus two 1″ x 4″ pine straps to secure in place with through bolts and lag screws. Get ready to stand in the stream for a while with your waders on with a brace and bit in your fist and a ratchet wrench in your back pocket.

The twenty-four inch 1" x 4" strap is fastened with four ³⁄₈" x 2½" lag screws and washers.

One of the main trusses intersects with a beam. Note that 3½" nails are used on the small right-hand side of the beam instead of bolts.

Ice and snow blanket the bridge making a path across the waters.

I put a couple of cement blocks in the stream to stand on so I'd have good leverage. An A-ladder would also work if the stream bed is solid enough.

There are two 4'3" (approx.) vertical struts that attach at their tops to the A of the trusses on each side and lead down to a pair of 5' horizontal braces that pass under the main beams across the center point of the bridge (see perspective drawing). At the apex of each pair of trusses the pine straps are through-bolted across the inside of the A using 4" x ⅜" galvanized bolts. The horizontal braces sandwich the vertical struts between them. Their 2" side is uppermost, touching the main beams. The bottoms of the vertical struts are attached to the ends of the horizontal braces with two 4" x ⅜" lag screws, one on each side. HELPFUL HINT: The easiest way is to get two bar clamps long enough to reach all the way under the bridge. Ask a helper (with boots) to clamp the braces and struts together on one side while you clamp them on yours. Then you can drill through one set, move across and drill the other set (*then* dismiss the helper, in case the clamps cut loose).

I drilled pilot holes, then drove in the screws with a ratchet wrench. If you have a cordless drill, you could bore the hole to through-bolt these three members with a 9" x ½" galvanized machine bolt. For me, it was just too difficult to bore through 8" of oak with a brace.

I had been there for some time as it was, first on one side of the bridge, and then on the other. Before I could screw the lags into the bottom ends of the struts to attach them to the braces, I had to attach the tops of the two struts to the apex of the truss members by boring two holes with my auger and bit through 2" of oak strut and 2" of pine truss on each side. Heavy work, even though I was using the highest quality bit I could buy. You need it for oak.

The through bolts I used were galvanized 5" x ½". Everything carries washers, too, both ends of the machine bolts and the head ends of all four lag screws. They keep the nuts and screw heads from biting into the wood. They provide a larger area of purchase for a tighter, firmer joint.

Why didn't I drill all these holes with power equipment when I had access to electricity back up in the pre-fab area? There is no way of knowing exactly where every connecting point for these will fall until the main beams and truss members are actually in place. The principle behind these struts and braces is that the vertical struts give additional support to the diagonal thrust of the truss members. Any downward pressure on the bridge is also transferred to the vertical struts and distributed at their bases to the horizontal braces that run beneath the bridge. Somehow it all works together.

WALKWAY

Finally, the walkway treads must be nailed on. I cut them with about a 5" overhang on each side. I left the ends of these boards uncovered, but it would also look nice to cap all along the ends. A 2" x 2" or 2" x 3" pine cap could very nicely go along the outside of the trusses too. I cut thirty-two 5′ lengths of 2″ x 6″ pine for the treads. Don't forget that some of the treads will be cut shorter to fit inside the space where short lengths of walkway pieces to accommodate the truss and vertical braces go. Using two nails per end, I just started nailing at one end and worked my way across using a 1″ piece of wood as my spacer. I had to check each one before nailing to make sure I was still at right angles to the beams because rough wood varies a lot. I also had to check constantly to see that I was leaving the same amount of space on either side of the over-hang. For me, it was good insurance to travel through this task with a 2′ piece of two-by-four along with me to line up the ends on either side and a 5″ mark on it to test the overlap. If you want real perfection, and certainly if you are going to put a cap on the ends, you should cut all your pieces a few inches longer than needed and then, after they are all nailed down, take a chalk line and snap a line from one end of the bridge to the other on both sides so you can cut a neat straight end on all the walkways. This is best done, of course, with a hand power saw, but if electricity is not available, a good old hand saw will do fine—just takes a little longer.

You may want a ramp or step down at the ends of the bridge, depending on the bank and the height of the walkway. My bridge will hold a lot of weight; it's strong enough to drive my small tractor over it, in fact. If this bridge were to accommodate more traffic and needed to be wider for a heavier vehicle, I would certainly put a third 4″ x 8″ oak beam down the middle, sitting on its own piers at each end.

Happy crossing and no tolls or trolls.

The triangles and straight lines of the bridge fit into their setting without standing out too much.

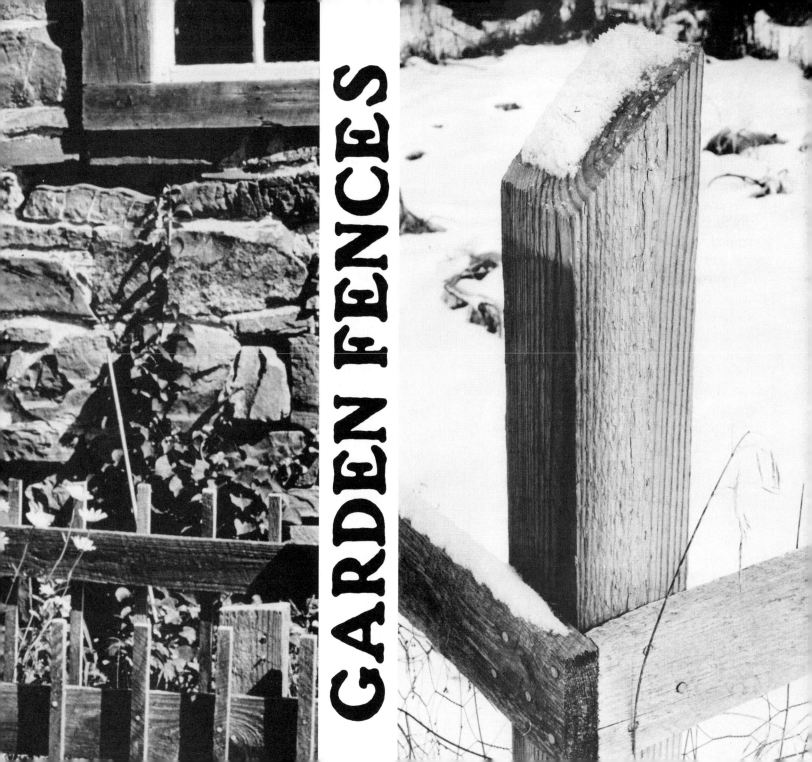

GARDEN FENCES

Fences have many uses—keeping animals in or out, marking a garden's dimensions, and hiding things that are better off hidden, like garbage pails or compost heaps. They function architecturally to define perimeters. In a sense, a fence is the frame around a picture. With so many beautiful fences available, it's hard to believe there are so many fences that are ugly and ruin the areas they surround. A fence is an important environmental detail that must be considered very seriously when introducing it to the general scheme of your property.

Before I go any further, I want to extoll the virtues of the post hole digger. For those who do not know this tool, a drawing is provided. I cannot tell you how important this tool has become in my life. Generally it's all that's needed to dig a hole big enough for a 4″ x 4″ post. Its cutting edge bites into the top surface of grass roots and then proceeds to chop its way south without making a large opening at the top. In order to use it, first hold the two handles straight up and parallel; then thrust down, twisting a little. Spread the handles apart and pull up some dirt. It is really a mini steam shovel (your steam). Now I know some people would rather just grab a shovel and start digging, but this method does not work for post holes. The shovel's hole is too wide for it's depth, so the sides are soft. To get the dirt out of the bottom, the sides must be widened, then filled in with loose earth around the poles. If at first the post hole digger seems unlikely to do the job, stick with it and you will be surprised how fast you can dig to a depth of 24″ to 36″. It is also a great exerciser for your arms as it chips through most stones and loosens the big ones enough for you to remove with your hands.

Another thing to consider is the best time of year to dig post holes. Generally, in early spring the earth is nice and soft from the spring rains. This is the ideal time to dig your post holes if you can, because later the earth loses its moisture and gets very hard, and you'll feel like you're on a chain gang in some prison. Early in the morning is also a good time, because it's cool and you will have more strength after a night's sleep. Plan to dig just two or three holes in one session. Then go back to the job later in the day or the next morning.

RUSTIC PICKET FENCE

Sometimes it's necessary to experiment with a design. The height of a fence, for instance, can make a significant difference. The low rustic fence shown at the beginning of this project started out a lot higher, but I realized that I wanted my zinnias to stick up over the fence as if contained in a sort of backyard vase. The herbs at the other end also grew to a size about equal to the height of the fence and looked lush rather than swamped or sunken in behind bars.

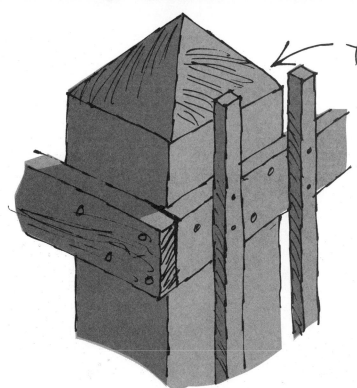

the four-sided post top as an alternate idea to the flat top.

irregular tops. For a more formal approach, the tops could have had four-sided peaks, as the illustration shows. The pickets were cut from 1″ boards into one-by-ones 21″ long. The crosspieces are one-by-fours on top and one-by-twos at the bottom. I used a 4″ spacer stick as I nailed along and kept a sort of jaunty up and down feeling to establish its air of rustic informality. Of course I checked the first picket with a level to be sure it was properly vertical. (For a neater, more formal fence, you can nail a section of pickets to the crosspieces on a workbench where you can control them, and then nail the crosspiece sections to the posts.)

The location and role of this particular fence defined its need to be sturdy but informal— blending in with a stone house and acting as a controlling factor for the plants cultivated in an uncultivated surrounding. (The stone house is an old summer kitchen that, in the old days before air conditioning, was used for cooking, baking and making candles during the hot summer months.) To build the fence, I first sank six 4″ x 4″ pressure-treated posts, then cut off flat but slightly

A steam shovel in its own time — the famous post hole digger.

127

I was going to stain the pickets dark brown but didn't get to it right away; now I love the way they have weathered and have left them as they are. A cap top was originally in my thoughts, but the horizontal line would have ruined the soft effect. A 45-degree cut on the picket tops would also be striking. (Next time.) Zinnias, do your thing!

Let me make another point; you should feel free to change your mind when designing. Sometimes a thing just does not look right, even though you planned it carefully ahead. If so, stop and change your route—you'll feel better for it.

GARDEN PLANK FENCE

This is a very simple country fence with 4″ x 4″ posts and either one-by-fours or one-by-sixes in two rows all around. The tops of the posts are slanted at a sharp angle, which allows them to shed water, of course, but also lends them an architectural style pleasing to the eye. The posts are 7½′ apart to accommodate the stock 8′ plank with the top plank about 33″ off the ground. The posts are approximately 48″ above soil level on the high side of the slope. (More about sloping land shortly.) It is a good fence to line with

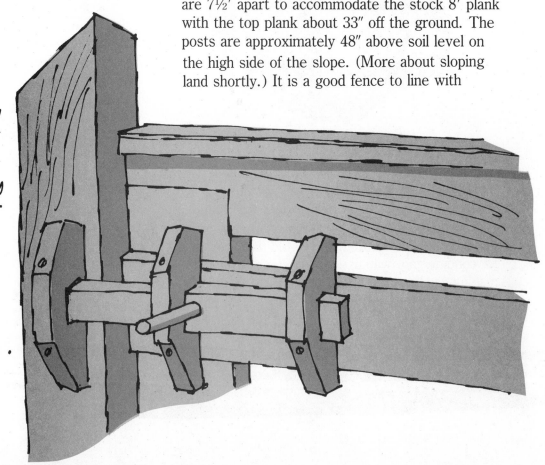

A ¾″ hardwood dowel and some scraps of oak or pine and you have something very authentic.

128

A garden plank fence stands out in bold design against the snow.

chicken wire (galvanized please) to keep the bunnies and woodchucks out. If you think that your woodchuck will ignore your fence and dig a hole under it into your garden, you may have to dig a trench and bury the bottom of the wire down in it, or resort to the old shotgun. The first year I had the garden I put the wire up only 28″ high all around. Then one day I went to pick some vegetables, only to discover a young woodchuck staring at me as he sampled my lettuce. He very grudgingly hopped out, but not until I prodded him a little with a rake. They are very stubborn guys and will sometimes stand their ground and defy you to do anything about it. The next spring I added another layer of chicken wire.

The interesting part of this plank fence is the gate. I had fun building the sliding bar latch. The drawing shows the shape of the latch parts, and

A closeup photo gives the tactile feeling of timber wood and also details of the sliding latch.

the metal hinges are a type easily obtained in most country hardware stores. It's best to bolt the hinges onto the gate rather than to use lag screws that become loose in time. The drop-on type hinges have a pintle with a screw thread, and the post must be predrilled to accept it. A wrench is needed to turn this pintle screw into the wood. One trick when hanging a gate or door is to put a piece of wood under the gate which represents the space that will eventually clear the gate from the ground. While resting the gate on this piece of wood you can locate exactly where the pintles should be screwed in. The strap part of the hinge should already be attached to the gate.

When constructing, mortise out the horizontal top and bottom crosspieces so that you can nail more effectively into the vertical sides of the gate. Pegs would be nicer. Drive dowels through with glue, two to a corner. The top plank was capped and finishes the look of the gate. The diagonal "Z" brace adds strength and keeps the gate from sagging. This brace also continues the horizontal lines made by the planks, because the gate rises above the height of the planks (see the picture—the effect is hard to explain in words). Wood hinges made of oak and with hardwood pegs for the pintles could also be used.

A question that frequently arises is how to cope with sloping land when building a fence. In my case, the garden sloped a little bit. When fencing a small area, it is best to keep the horizontal planks horizontal rather than slanting them with the land. This means that the posts on the lower side of the slope will be somewhat taller than those on the higher side.

When should you cut the posts? Start with the corner posts, as follows. First determine the height you want the top of the post to measure from the ground on the *higher* side of the slope. Dig a 24″ hole and add 24″ to the pole height you have decided on. Cut the top at a slant and drop the post into the hole. Next dig the hole for the corner post on the *low* side of the slope. Take the second post, uncut, and drop it in the hole. Run a level line from post to post with a line level (a small level that hangs on a string) and then you can tell exactly at what height to cut the second post. Cutting the rest of the posts is easy!

Don't forget to add some gravel or small stones to the bottom of the hole for drainage. If your land slopes a little more than my garden did, you may want to put a third row of planks on the lower side. This horizontal business is not a hard and fast rule though; the situation must always be taken into account. In northern New York State where our land is are beautiful horse farms; they have miles of plank fencing four and five planks high that follows the terrain religiously, but of course the area is vast and their meandering lines are quite beautiful to behold. (See photo.)

Using pressure-treated posts, because they will be in direct contact with the earth, is a good idea. If they are untreated, soak them in creosote. The planking material does not have to be pressure-treated since it is free to dry in the air. The pressure-treated wood is almost twice as

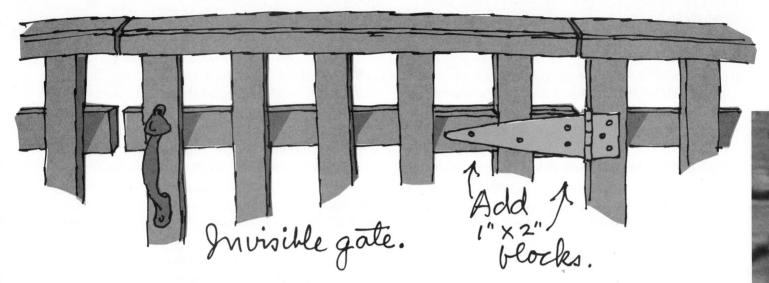

Invisible gate.

Add 1" x 2" blocks.

expensive as regular lumber, and I have not seen any proof positive that it lasts longer. Any stained or painted wood is going to hold up much better than untreated wood, but I do not feel that pressure-treated wood is a substitute for painted or stained, and certainly not at the price. Also, I prefer stain rather than paint for wood exposed to the weather, so it can breathe and dry more thoroughly after being rained upon. Paint does not breathe. It covers, and moisture creeps under it and blisters it off all too frequently. Dark brown stain is very handsome; gray gives the wood a velvety, weathered look. As you should in all outdoor building projects, use galvanized nails, either common or finishing, because plain steel will rust and streak the finish.

PICKET FENCE WITH CAP

This fence looks nice running along a sidewalk at the edge of a property and could have an arched entrance with a gate. The pickets can be close enough to hide garbage pails and to keep raccoons and other animals out of the area inside. Our fence hides not only the garbage but also two propane tanks of the ugly persuasion that adorn so many country houses. Luckily, what I had thought was a weed one year turned out the next to be a large Hollyhock growing right next to my fence. When I let it grow, it blessed us with beautiful red blossoms all summer long. It's intriguing how flowers and bushes play an important role with fences. Privet hedges look handsome packed thick behind this kind of fence, especially when they are trimmed at a height about equal to that of the cap.

The cap picket fence is made the same way as the rustic fence, with two 1" x 2" crosspieces. The pickets themselves are 36" high by 2" wide and are cut out of 1" stock. I left them rough, but smooth lumber works well too. The cap is

1½″ wide and given a beveled edge. I cut an invisible (so-called) gate into my fence. It may seem a bit tricky, but I did not want a fancy gate in this case. Nantucket is one place where this type of fence is common, and quite often the gate is hidden, as it were, giving the observer the impression of an unrelieved horizontal line. Houses in Colonial towns were quite often near the sidewalk, and I imagine the builder wanted the front door to be the center of interest and not the gate. (I deduce this—I don't claim it as a known fact.)

Hollyhocks enhance the capped picket fence attached to our home.

I wish I could claim to have built all the fences pictured here, but it was fun to seek out some very interesting designs that other people have created in the hope that they will inspire you as they did me.

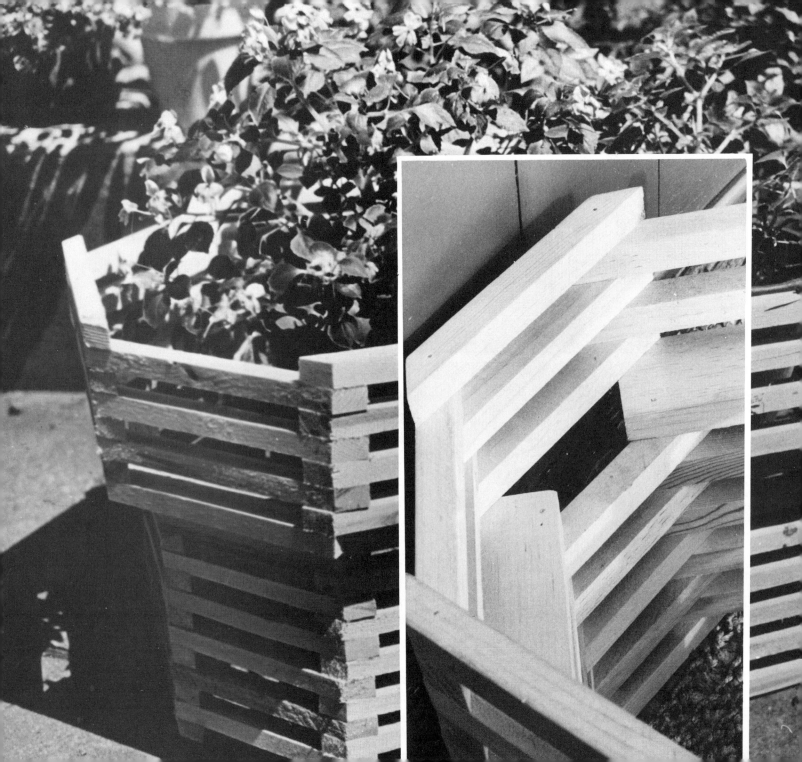

OCTAGON PLANTERS

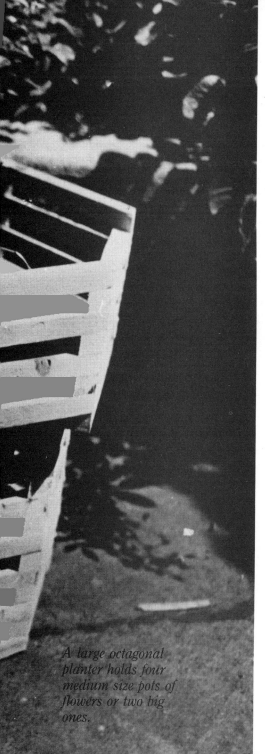

Wood scraps and the accumulation of many more wood scraps over the years gave me the idea for this planter. I got so excited about this project and the look of my first finished planter with a plant in it that I made a large one to hold several plants. At one point I even began to consider the idea of going into production and selling them at craft fairs, and so on into the sunset. But after making about six or seven, I realized that I was not the production-line type, and after cutting a couple hundred of these little pieces of wood, another problem became apparent. The repetition of cutting piece after piece can lead to a dangerous slip at the table saw. It's one thing to turn out an item for your own use or maybe one for your Aunt Minnie to whom you owe the world, but mass production for the masses has its drawbacks.

The repetitiousness of the task did stimulate me to design a jig to make the work go faster when I cut all those identical short pieces (see illustration.) I'm no professional jig maker, but I learned to appreciate fine woodworking when I realized jig making was at the root of this country's well-known productivity—because it offered

A large octagonal planter holds four medium size pots of flowers or two big ones.

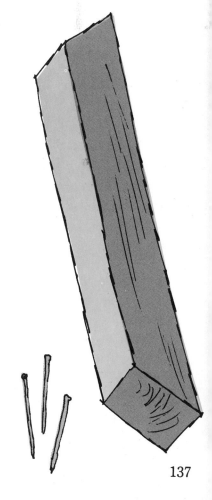

a simpler way to produce more items to meet so many people's needs. My crude little device enabled me to cut lots of pieces without marking each one in the same place with a pencil. It is adjustable for different lengths of stock except very long ones. The thin member of the jig underneath should be made to fit the trough of your table saw so you can slide the jig back and forth. The flathead bolt and wing nut allows you to adjust the jig for distance from the blade.

BASIC ROUND PLANTER

All angles on the pieces are cut at 45 degrees. For the round planter, you will need the following pieces—measured between the longest points.

> For the 5 top rows: (20) 7¼″
> (rectangular dividers) (4) 6″ x 2 ¼″
> For the 9 bottom rows: (36) 6″

The dimensions could apply to scrap stock of any uniform thickness but I chose ¾″ x ¾″. It could also be either rough or smooth-finished, whichever pleases your eye. I have not experimented with exotic woods, but the idea is exciting. The design possibilities are almost limitless, and you may decide to create a variation on mine. It's good to draw your idea out on paper first, though, to determine the sizes you need to cut.

Use thin, galvanized finishing nails just a bit shorter than the thickness of the two pieces of wood together. In this planter I used 1¼″ nails. Softwood will almost always take a thin nail without splitting, but for a hardwood model you should predrill the holes with a bit that is slightly smaller than the nail. Gluing is a good idea, although I found that my unglued ones are quite sturdy.

The first nine rows are made from the 36 short pieces, so each row has only four pieces. Put one nail at each joint and position the two ends together by lining up the 45-degree angles. Center the nails in the first row, and in the next offset them a bit from the center so that the nail you're about to drive doesn't hit the nail below. SUGGESTION: Sand the first nine rows before going on with the rest of the rows. Be sure to clamp the work down.

When nine rows are nailed you are ready to assemble the four rectangular dividers that make up the tenth row, following the same pattern. But before nailing these dividers, a little trick is involved. Cut a piece of two-by-four the height of the first nine rows. Stand this under each corner as you nail to give you the support needed to drive the nails on the five upper courses. For the remaining five rows, use the twenty 7¼″ pieces in alternate rows.

45° jig for angle cuts.

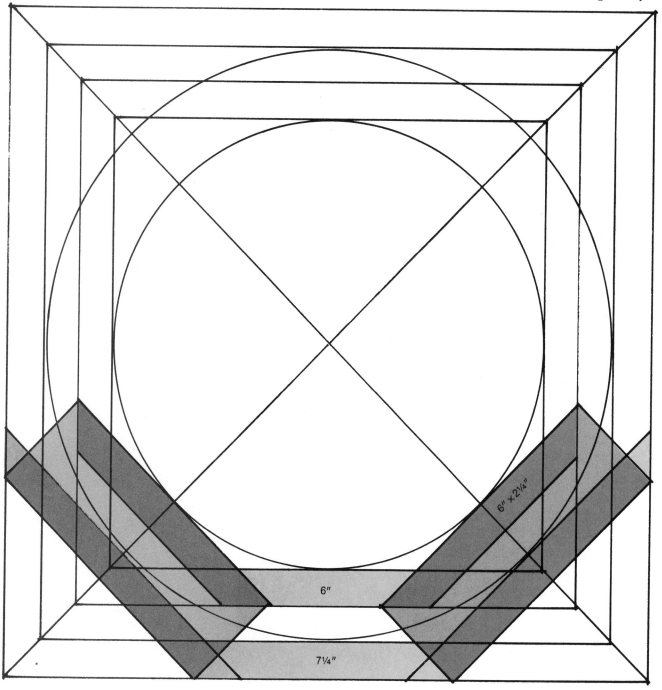

Here is one method of arriving at dimensions for the planter. The possibilities are endless. The drawing is half size.

6" x 2¼"

6"

7¼"

139

For a smooth planter, use kiln-dried preplaned lumber. It also can be bought ¾″ thick so it needs only to be ripped on the table saw. Use either a hand belt sander or a bench sander on each side to smooth the 45-degree edges which will generally be a little rough.

It's a lot of fun to put one of these together, and if you follow the sizes given here, the collar of a 10″ clay pot will sit very nicely on the four rectangular dividers with its bottom slightly off the ground.

OBLONG VERSION

The oblong planter is a little more complicated. Four different lengths are involved and the usual four rectangular pieces. To achieve the most pleasing proportion, use thirteen rows on the bottom, then the rectangular row, topped by seven final rows.

```
          For the 7 top rows: (8) 10¼″
                              (12) 10″
                               (8) 22¼″
(rectangular dividers) (4) 8⅜″ x 2″
       For the 13 bottom rows: (24) 8⅜″
                              (14) 9½″
                              (14) 21½″
```

Build this planter following the directions for the basic round planter, but here you will need a two-by-four that is a little longer to support the divider so you can nail the upper rows. These planters look great on wood decks because they repeat the striped texture of the deck's planking.

This macro view shows the interlocking construction of the planters.

A pair of octagonal planters sporting their plumes.

DROP LEAF HARVEST TABLE

In our home we have some interesting old things including a modest-sized drop leaf table that is used as a dining room table but, at best, seats only six people. The room it is in could take a longer table and make holidays like Thanksgiving more festive gatherings. So I was determined to make a larger table, Colonial in feeling, with drop leaves that could be put down when not in use and giving us the option of raising only one leaf for normal use.

SELECTING STOCK

This project calls for air-dried full-dimension timber, but it must be planed and sanded to produce a finished piece of furniture. There is no reason, except cost, not to buy finished pine stock at a lumber yard or look for hardwoods like cherry, which I think is particularly beautiful, or maple that traditionally have been used for making classic furniture. Some lumber companies sell by mail and will ship wood to you if you can't get there yourself. They usually advertise in woodworking magazines and also in *Wooden Boat* magazine.

With rough wood, always select boards that are longer than you will need because it is some-times hard to see what the wood is like until it is planed. I also buy a few extra boards so that during the drying time if a piece turns out bad some-how, I still have others.

For an 8′ table, for example, I selected some 12″ pine boards I had bought a while ago and dried and cut them into 9′ lengths. Then after planing, I cut it again to the proper length. When picking lumber for furniture, be very careful about knots and grain. Loose knots are ones that look very black and crumbly. They are usually clearly sharp-edged and look as if a hammer blow would knock them out. Knots that blend into the surrounding grain and are lighter brown in color are better bets and look beautiful when refinished. Some cabinet makers work only with clear wood, but the price is very high and the finished piece looks more formal. For the same money I would take cherry rather than clear pine.

A few words on grain. Generally the best boards to try to find are those where the grain is more vertical through the thickness of the board rather than slanting in a curved effect. The curved lines are the annual rings that mark the tree's years of life. The best way to see this is to pick up the end of the board and study the picture that the end grain will show you. If it is too rough to read, sand down the end smooth. When the grain lines appear vertical or almost vertical, it will make ideal stock for table planks because it is less likely to cup or warp. A cross-section drawing of a log shows where different pieces of wood were cut from the log: boards with a vertical grain (A) and boards where the end grain curves or arcs (B). Lumber with a curved end grain gives you a very interesting grain on top, but you pay for that with its tendency to cup or warp along its length.

You will find more curved-grain than straight-grain in a given stack at your lumber yard because of the way lumber is milled today. Years ago when lumber was plentiful and cheap, they cut a log in such a way as to get all vertical grain. Then that was felt to be too wasteful. Per-haps, but I think the reason is it took more time to cut it that way and we sometimes tend to do things the easy way today. The old way of cutting was called "quarter sawn," because the timbers were cut at right angles to the heart or center of the tree (C). If you prefer the look of the top grain on boards with curved end grain, be aware that it might cup. A precaution to take with any lumber is to stack it properly. In either case, rough timber or kiln-dried lumber, the stuff must be stacked on stickers and kept away from high heat. Lumber with a curved grain will also fare better while drying if the concave side of the curve faces down. If you are using rough timber about an inch thick, let it dry for at least 2 months. In making a piece of furniture, the inside of the curve should face whatever it is screwed or glued to. If, after planing, a board shows signs of cupping before you can use it, clamp it to the

Boards today are cut all in one direction through the log.

bench with a C-clamp and a cross brace of wood at each end. Wood ½″ and ¾″ thick tends to cup more than 1½″ thicknesses.

One of the dimensions you don't get from a lumber yard is the full 1″ thick board. They usually have ¾″, ⁵⁄₄″ and 1½″ thicknesses. The ⁵⁄₄″ should be 1¼″ thick but often is planed down to 1⅛″. Wood ¾″ is thick enough for a table top, but I prefer ⅞″ or ¹⁵⁄₁₆″, especially for a big table with large legs. Rough timber comes a full 1″ thick and can be planed down to the thickness you prefer. The ⁵⁄₄″ lumber-yard lumber can also be planed down but is more expensive. The un-

PLANING

I have a small bench planer that I use for rough wood. If you haven't invested in power equipment, use a hand plane and then a belt sander followed by hand sanding with a block. A power hand planer is another alternative. You can probably still find one for less than $100 if you look in the catalogs published by tool mail-order houses around the country. This tool is useful and versatile, but on a wide board it will leave small ridges as you plane no matter how carefully you work. Use a very shallow blade setting so the ridges can be erased with a belt sander afterward.

derside of the table top does not have to be finished as smooth as the top. The old timers felt that way too, I have observed, from feeling under table tops. If you like to use a router and want to give the table edge a molding, the wood will look thinner and more delicate.

THE TABLE TOP

The boards I selected for this project were 1″ by about 12″ wide. The table top is made up of two boards and totals 23⅝″ in width when trimmed and planed, not including the leaves. Each drop leaf is about 11¼″ in width. First plane all surfaces and cut to the 8′ lengths. Decide how the grain of the four boards will match best by laying them side by side. Make sure the two edges that will butt fit nice and snug without any gaps. Plane the edges a little more if necessary. To glue and dowel the two boards together, place the best-

Lining up the dowel holes for drilling.

matching pair of boards in a bench vise with the two edges to be butted facing up. Locate the dowels about every 12″ to 18″ apart with two end dowels about 2″ or 3″ in from the ends. With a right angle, draw a pencil line across both boards at the points you want the dowels.

Purchase a dowel guide for your drill so the holes will be straight down. This is a marvelous tool whose accuracy comes in handy many times. The most common type, pictured in the section on tools, will handle holes up to ½″; larger ones are available for projects with bigger holes. Some guides will even line up the holes in both boards at once for drilling. The drill guide when positioned over the board and tightened to fit the thickness of the plank will automatically center itself. Just line it up with your pencil line (use a flashlight to check this) and clamp it tightly, but don't mar the surface. Select the appropriate size bit and drill away, making sure to go a little deeper than half the length of the dowel. There is nothing worse than drilling the hole too short and later finding that out after you have put glue on the edges and go to clamp the whole thing together. For short holes such as these, the work will go faster if you take the time to measure on your drill how far the bit should go in and tape its circumference at the point where you want to stop drilling. Don't forget to add on the height of the drill guide. I used 5/16″ precut dowels which can be found ready-made in several diameters in tool catalogs and lumber yards. Of course, you can cut your own dowel pegs from long hardwood dowels, but the precut and boxed ones are

fluted along their lengths which, supposedly, lets them hold more glue and make a stronger bond.

GLUING

You need at least two bar clamps. For a long table such as this, three or four clamps will more equally distribute the pressure all along its full length (I used four). Put in all the dowels with glue and then coat both facing surfaces with glue, spreading it over the surface, and clamp. Wipe off excess glue with a wet rag. There are several types of wood glues. Some are already liquid in form, others called plastic resin glue are a powder to which you add water. This is the type I favor. When you mix this type, follow the directions on the can faithfully and don't use it in temperatures below 66 to 70 degrees. This makes a fine lasting bond and is used a lot by boatbuilders as it is water resistant.

Sand the top surface after the glue dries using #80 and then #120 grit paper. A belt sander saves lots of time. Then, by hand with a sanding block, bevel the sharp edges all around very slightly. If you sand the end grain down very smooth, the stain will not soak in as much and make the end darker than the grain on the table top.

LEGS

The actual measurement of the table base is less than that of the top, of course, but the amount can vary depending on the style you like. I left a 1¼″ overhang all around. With this overhang, the base will measure 93″ x 21½″ including the corners of the legs. Since the average comfortable table height for eating and working is about 29″ (it was lower in Colonial times because people were shorter and chairs lower), the legs then will be 28″. You can design your own shape for the legs. There are no fixed rules except what looks right to your eye. If you can draw, render different shapes to scale on paper first to give you a better idea of how the one you think you want will look. I favored a four-sided Sheraton type leg that tapers down to about 2″ square at the foot. I started with 3½″ x 3½″ stock and began to cut the taper about 6″ down from the top. I used a band saw and planed the sides with a hand power plane. One thing the power does better than a hand plane is to slice right through knots without leaving rough edges or holes. Finish by hand using #120 grit around a sanding block, slightly rounding the edges.

CUTTING MORTISES

Refer to the drawing for the best understanding of the mortise and tenon type of joint where the sides attach to the legs. I used 1″ stock for the side pieces on this table; the shorter end pieces measure 21½″ and the long side pieces 93″. The end pieces are notched to make a ½″ x ⅞″ tenon. The tenon is on the outward face of the side pieces for a neat appearance. Cut the ½″ dimension across the grain with a back saw, then chisel away the ¾″ residue. The short ends can be cut with the dado in one sweep set at ½″, if you prefer.

The mortise in the leg can be cut best as fol-

the mortise cut in the leg. →

the tenon fits into the mortise. ↘

lows: put a dado blade in your table saw and set it for ½″ width and the proper distance you have chosen from the table saw fence (see measurements in illustration). Once this is right, use a 3½″ x 3½″ scrap to test your measurements. Then all eight cuts can be made. Since the cut is 5″ long, you will have to mark the top of your table saw with tape to show where to stop. This trick gives you the accuracy you need. When these cuts are made there will still be a rounded end inside the mortise because (you guessed?) the saw blade is round. This will have to be chiseled out using a ½″ chisel and mallet on all eight cuts.

ASSEMBLY

Join the legs and short end pieces first. Clamp and glue one end, then, when the glue is set, clamp and glue the other end. Before the two long sides can be assembled, the four cut-outs for the leaf braces must be cut into them. These cuts are about 20″ long and their ends are cut at a 45-degree angle (as illustrated). This acts as a stop when closing the braces into the sides of the base. It is easier to cut these on the bench before gluing and clamping them to both ends. Because this table was longer than my longest bar clamp, I had to use the cut-out for the braces as a clamping point at one end. Four bar clamps will be needed to glue both sides at the same time.

After all four legs and sides are glued and before the top is put on, mark and drill dowel holes. For this type of visible dowel, use ¼″ hardwood dowel as sold in lumber yards. Be sure you drill through both the leg and the tenon. Put two dowels on each leg side, measuring, so they all will be in the same positions. Cut them a little longer than needed, put glue on them, and ham-

Four braces, each 20″ long support the leaves. Note the 45° angle cut.

the 1"x1"x20" brace in position.

Table assembled and waiting for the top.

"Forstner" bit hole.

mer them in. Slice the tops flush with a hacksaw and then belt sand to finish flush. The 1″ x 1″ braces are 20″ long with the ends cut 45 degrees and are secured with a piece of ¼″ metal rod driven through the brace into the side. This hinges the brace and you can swing it back and forth into position. The angle cut allows the brace to fit back snugly when not being used.

149

To assemble such a large table top, you will need two horses, each long enough to support the whole top plus one leaf. Pad the horses with towels so while you are working you don't dent the surface of the table. To screw the top and frame together, position the frame upside down on the top and clamp it in the desired position before drilling holes. Attaching the frame to the top use a ¾″ Forstner bit that can make slanted or angled holes, then use a regular countersink bit drill for the screwhole. As the accompanying drawing indicates, you can angle the screw into the table top from the inside of the frame. Put four screws in each side. I used 2″ screws, but this can vary depending on where the Forstner holes land. You must calculate how much distance remains to go without going through the table top and select screw lengths accordingly.

To attach the leaves, position one leaf, topside down, on the horses, lining it up next to the inverted table top. A wide hinge is better than a narrow cabinet type because the leaf has a lot of downward thrust. People sometimes accidentally let a leaf drop a little too fast and it hits the legs—everyone has done it. A short cabinet hinge might not be able to take it and the wood could split right where the hinge is attached. This table has three pairs of 1¾″ x 2⅝″ brass hinges positioned in the center and 7″ in from each end. (I found them at Garret Wade Tools in New York City, where they have a terrific mail-order catalog of tools and hardware, which is an education to read by itself.) Since the hinges will be flush-mounted, open one hinge and use it as a pattern, drawing its contour on both the leaf and the table

The table ends and leg make a neat appearance because the tongue of the tenon goes directly into the mortise and does not stop at the leg.

150

*The completed table
shown from one end.*

top—in all three locations. Chisel this area out just shallow enough so that the hinge will lie flush to the surface. Screw the first three hinges in place and proceed to the other leaf. If your horses are not wide enough to hold the top with both leaves open (almost 4′), fold up the first leaf and tie it to the legs while hinging the other side. You are going to have to tie both leaves to the legs anyhow in order to turn the table right side up. It's heavy, so get help.

FINISHING

There are metal sliders available that are very shallow in height that can be put on the bottom of each leg. A modernism, perhaps, but it does save wear on the bottoms of the legs if the table must be moved frequently.

The finish for the pine starts with a color stain of your choice—one to two coats. Then apply two or three coats of a natural varnish with fine sanding lightly between coats using a #120 grit. Cherry, to my taste, needs no stain because it has its own color. Do not use a shellac finish because it will turn color if you put hot or very cold dishes on it. Minwax has an oil finish that is very hardy. It comes clear and walnut in color and is best rubbed on with a cloth. Use a fine steel wool between coats.

All that remains is to plan the christening feast. Happy dining!

Use brass hinges like these—square rather than long and narrow—for flush mounting.

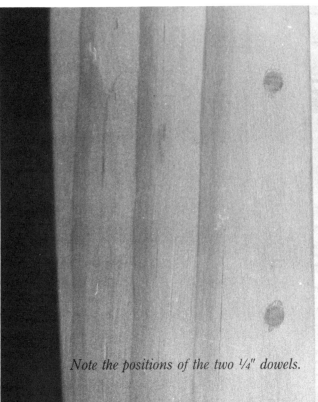

Note the positions of the two ¼″ dowels.